An Illustrated History

The Evolution of AERIAL FIRE APPARATUS

Walter M. P. McCall

Iconografix

Iconografix
PO Box 446
Hudson, Wisconsin 54016 USA

Library of Congress Control Number: 2009921083

ISBN-13: 978-1-58388-235-1
ISBN-10: 1-58388-235-9

09 10 11 12 13 14 6 5 4 3 2 1

Printed in China

Cover and book design by Dan Perry

Copyedited by Andy Lindberg

On the cover:
Upper left:1917 Seagrave 65-foot Water Tower, St. Paul, Minnesota. *Walt McCall*

Upper right: 1930 Seagrave 85-foot Tractor-Drawn Aerial, ex-Syracuse, New York. *Walt McCall*

Middle left: 1946 American-LaFrance JOX 85-foot Service Aerial, ex-Rensselaer, New York. *Walt McCall*

Middle right: 1970 Crown Firecoach 65-foot Snorkel, ex-Rohnert Park, California. *Walt McCall*

Lower left: 1977 Sutphen 85-foot Aerial Tower, Moorhead, Minnesota. *Dave Stewardson*

TABLE OF CONTENTS

ACKNOWLEDGEMENTS

The majority of the photographs in this book came from the private collections of the author, and that of longtime friend and fellow fire apparatus historian Matthew Lee. Yet again, Matt opened his rows of filing cabinets and stacks of photos and sales literature to us, for which we're deeply grateful. We simply could *not* have done this book – and various others – without unfettered access to Matt's truly comprehensive collection and his unfailingly constructive advice.

We also extend our sincere gratitude to a number of other individuals who assisted us in this project. Thanks to *Dan Jasina, Dave Stewardson, Steve Loftin, Ron Heal, Chuck Madderom, Bill Noonan, Neal Van Deusen, Steve Hagy, Shaun Ryan, Dan G. Martin, Bill Friedrich and Kent Parrish.* If we've inadvertently overlooked anyone, the fault is entirely our own.

In the interests of brevity, we've employed some initial codes in the photo credits. *MLC* indicates a photo from the *Matthew Lee Collection*, and *WMC,* the author's. *JFS/WMC* indicates a *John F. Sytsma* photo from our own collection. All other photos are credited individually.

BIBLIOGRAPHY

American Fire Engines Since 1900 – By Walter M. P. McCall, Crestline Publishing, 1976

Illustrated Encyclopedia Of American Fire Engine Manufacturers – Walter M. P. McCall, Iconografix, 2007

100 Years of American LaFrance – Walter M. P. McCall, Iconografix, 2005

Elevating Platforms – Paul R. Barrett, Iconografix, 2006

Aerial Fire Trucks – Larry Shapiro, MBI Publishing, 1999

Hook & Ladders – Larry Shapiro, MBI Publishing, 2002

Seagrave 125th Anniversary History – By Matthew Lee, 2006

Pictorial History of the Fire Engine – Volumes I & II, 1997, 1999, By Matthew Lee

Wheels Of The Bravest 1865-1992 – John A. Calderone & Jack Lerch, Fire Apparatus Journal Publications, 1984, 1993.

History of the American Water Towers, Bill Hass, Kes-Print Inc., 1989

Enjine!-Enjine! – Quarterly publication of the Society for the Preservation & Appreciation of Motor Fire Apparatus in America, Inc.(SPAAMFAA).

FOREWORD

"The story of aerial ladder apparatus is not one of startling innovation, but rather a narrative of slow but steady development," observed noted fire service historian Clarence E. Meek in an article published in *Fire Engineering* – the nation's leading fire service trade journal – half a century ago.

Published in April 1958, Mr. Meek's treatise, *The Legend Of The Aerial Ladder,* ironically appeared only a few months prior to the surprise debut of one of the most revolutionary innovations in aerial firefighting in America – the elevating platform. With its remarkable aerial mobility, the Chicago Fire Department's ground-breaking *"Snorkel"* subsequently swept the North American fire service like a prairie fire – the first real alternative to the aerial ladder since the latter's introduction some 70 years earlier.

As Mr. Meek, an Honorary Battalion Chief of the New York Fire Department and FDNY Librarian noted, former New York volunteer fireman Daniel D. Hayes made the long voyage around Cape Horn in 1866 to deliver an Amoskeag steam fire engine to the San Francisco Fire Department. Hayes chose to stay in the City by the Bay, where he became a SFFD machinist. It was there that he designed America's first *successful* mechanically-raised turntable ladder, which he patented on February 23, 1868.

A flurry of other similar inventions soon followed, but Hayes' proven ladder led the way. Hayes later sold his patents to the LaFrance Fire Engine Company of Elmira, New York, which built many *LaFrance/Hayes* aerials in the late 1800s and early 1900s.The other two conspicuously successful early American aerial ladders were the *Babcock,* patented by the Fire Extinguisher Mfg. Co. of Chicago in 1884, and the *Dederick* (1895) made by Gleason & Bailey in Seneca Falls, New York.

Other pioneer makes of aerial ladders in the United States included the *Skinner,* built by the George Hannis Co. of Chicago in the early 1870s; the *Scott-Uda,* an Italian-designed ladder made in New York ca. 1875; the *Gillespie,* made by the Warwick Machine Co. in Warwick, Rhode Island, and the *Arrow* turntable ladder made by the Michigan Fire Ladder & Engine Co. of Grand Rapids. In 1888, *E.B. Preston* of Chicago delivered a three-section, 90-foot *steel* aerial to St. Paul, Minnesota.

Another distinctly different type of aerial firefighting device which appeared in the late 1800s was the *Water Tower* – a portable standpipe designed for the sole purpose of lobbing a heavy fire stream into the upper floors of burning buildings. These are covered in a separate chapter in this book.

As the twenteth century began, horse-drawn *Hayes, Babcock* and *Dederick* aerials dominated the scene. The next advance in aerial ladder design was the development of the *spring-raise hoist* by Seagrave in 1902, followed by American-LaFrance's *American Automatic* spring hoist ca. 1903-'04. A pair of powerful coil springs took the muscle work out of raising the heavy two-section wooden aerial ladder from its bed.

But a far more significant technological revolution in the first decade of the twentieth century was the appearance of the first *motor-propelled* fire apparatus in 1905-'06. Seagrave exhibited the first motor-driven, tractor-drawn aerial ladder truck at a fire chief's convention in 1908. Within a few years, fire departments across the nation were replacing horses with two- and four-wheel auto tractors, the moist prolific of these being the *Christie* made by the Front Drive Motor Co. of Hoboken, New Jersey. In the 1920s, two newcomers to the burgeoning aerial ladder market eschewed the contemporary spring-raised aerial in favor of other lifting devices – Ahrens-Fox the *Dahill Pneumatic* hoist in 1923 and Mack, a power-takeoff *engine-driven* hoist introduced in 1929.

Despite the ravages of the Depression, the 1930s arguably saw more strides in aerial ladder development than in any other period in history. Peter Pirsch & Sons introduced the first *hydro-mechanical* hoist in 1931, followed a year later by the industry's first *service aerial* – a city service ladder truck equipped with a midship mounted "junior" aerial ladder. In 1935, Seagrave fielded the industry's first hydraulically-operated, all-steel aerial ladder. Pirsch delivered the first 100-foot, 100-percent powered aerial ladder truck to Melrose, Massachusetts, in 1935 (although *Babcock* is on record as having built a horse-drawn 100-foot aerial for St. Louis, Missouri, in 1892). American-LaFrance introduced its first all-steel, hydraulic aerials – with fixed tiller seat – in 1938, followed that same epochal year by its industry-leading JO/JOX *cab-forward* service aerials.

FWD delivered the last spring-raised wood aerials to New York and Chicago in 1955-'56. As alluded to earlier, the *elevating platform* made its dramatic entrance in 1958, sparking a revolution in aerial firefighting. The articulating type aerial platform that started it all was joined in the mid-1960s by the telescopic *aerial tower* pioneered by Sutphen, and Mack's competing *Aerialscope*. The 1960s saw its share of notable advances in aerial truck design. Presaged by industry leader American-LaFrance's trend-setting *700 Series* in 1947, the rest of the industry introduced their own cab-forward aerials in the 1950s and '60s.

Seagrave precipitated the next notable advance with its *Rear Admiral* in 1965. Although the rearmount aerial concept had been around since German manufacturer Magirus brought it to America in the mid-1920s, the compact, short-wheelbase rearmount aerial ladder truck was not widely accepted by the North American fire service until the late 1960s.

While the new rearmount aerials and aerial platforms were the industry standard by the 1970s, several supplemental aerial firefighting devices had been introduced by the makers of the original *Pitman Snorkel*. These included the *Squrt* – a 54-foot articulating boom master stream device (1968), and the spectacularly successful *TeleSqurt* – a 50-foot, two-section telescopic combination water tower/aerial ladder which has since been widely imitated.

The entire U.S. fire apparatus industry underwent an agonizing restructuring in the late 1980s and 1990s. Many of the great old names – Pirsch, Maxim, Mack and Hahn among them – faded away, succeeded by aggressive, innovative new aerial makers like *E-One* and *Pierce*. While the rearmount aerial ladder and platform reigned supreme, midship-mounted aerials and tillered, tractor-drawn aerial ladder trucks were making something of a comeback. As the first decade of the new millennium was winding down, more than a dozen companies were producing aerial ladders and elevating platforms in the United States.

More than 140 years after Daniel Hayes designed his *Combined Hook and Ladder Truck and Fire Escape,* the basic function of the aerial ladder remains unchanged – to facilitate rescues from upper reaches, and to provide a safe, stable platform to direct a powerful hose stream into the raging inferno from above.

Walter M. P. McCall
Windsor, Ontario
January 2, 2009

Boston's first aerial ladder truck was this ungainly looking 1876 *Scott-Uda,* photographed in front of the Washington Square Fire Station. Note *"Aerial Ladder No. 1"* over one of the station's doors. Three New York firemen were killed in the tragic collapse of a Scott-Uda aerial during a public demonstration of one of these Italian-designed ladders the previous year. *MLC*

CHAPTER ONE: MUSCLES & MANPOWER

Webster's Unabridged Dictionary defines the word *"aerial"* as meaning, *"high, lofty, reaching into the air."* That certainly was the basic design intent behind the invention of the first successful fire department-type aerial ladder in the United States some 140 years ago.

In the 1860s, buildings in America's burgeoning cities – office buildings, hotels and tenement houses – were growing ever-higher, making the rescue of people trapped on the upper floors of these structures in the event of a fire difficult, if not impossible. The longest ground ladders then available were massive 50- to 70-foot wooden extension ladders that weighed half a ton and required ten firemen to muscle them off the hook-and-ladder truck and tote them to the fire building. Even with *"tormentor"* poles, these long three-section *Bangor* ladders were bulky and awkward to position and raise. Once in place and fully extended, they would barely reach the fifth or sixth floors – certainly distressing for those on the seventh, eighth or ninth floors. Their only options at this point: a daring rescue by a fireman scaling floor-to-floor with a pompier ladder – or a leap into a life net or almost certain death.

America's first successful aerial ladder was patented by *Daniel D. Hayes* in 1868, with the first one delivered to the local San Francisco Fire Department circa 1872. Hayes later sold his patents to the *LaFrance Fire Engine Company* of Elmira, New York, which began manufacture and sale of the proven Hayes aerial in 1883. Here's an early *LaFrance/Hayes* aerial being tested in front of the LaFrance factory in the mid-1880s. *MLC*

A former New York City fireman, Daniel D. Hayes, was granted a patent for America's first successful self-supporting fire department-type aerial ladder in 1868. Hayes described his invention as a *"Combined Hook and Ladder Truck and Fire Escape"* – the term *aerial ladder* did not come into general use until some time later. The San Francisco Fire Department – where Hayes then worked as a foreman in the municipal shops – is believed to have placed the first Hayes-patent aerial ladder truck into service in the early 1870s, ca. 1872.

Hayes' practical design consisted of a long, two-section wooden extension ladder mounted directly to the hook and ladder truck's frame. The pivoting base of the ladder at the forward end of the apparatus was attached to a movable turntable, permitting the main ladder to be rotated 90 degrees or more to either side of the truck. The heavy ladder was manually raised by turning a large crank which had to be attached to the front of the apparatus – *after* the horses had been unharnessed and led away. The turntable was also rotated by hand. As is still very much the case today, precise "spotting" of the aerial truck in front of the fire building was imperative if the ladder was to reach a particular floor or window to effect a rescue or to get a hose line into the structure to attack the fire.

A true mechanical marvel at the time, the Hayes aerial marked a major technological advance in firefighting in America – one that would not really be equaled until the introduction of the aerial platform nearly a century later.

As the Twentieth Century began, the horse-drawn *aerial hook and ladder truck* had long ago reached maturity. Heavy horse-drawn aerial trucks had been in general use across the country for the better part of 30 years. Every metropolitan fire department in America had at least one aerial hook and ladder truck on its equipment roster. Most big cities also had a *water tower* – another distinctly different type of aerial firefighting apparatus introduced at about the same time as the aerial ladder.

The typical aerial ladder truck of 1900 was drawn by three horses (sometimes two, or four) and had a 65-, 75- or 85-foot main ladder. Perched far at the rear, a *tillerman* steered the long aerial truck's rear wheels. Other members of the truck company rode on the wide running boards, clinging to side-mounted ladders as the big truck hurtled through the streets to the fire. The largest piece of apparatus in the fire department, the aerial truck also carried an assortment of portable ground ladders, a life net and various other tools such as ropes, pike poles, cellar nozzles and battering rams.

The majority of the horse-drawn aerial ladder trucks in service in the late 1800s and early 1900s were made by one of three prominent American fire apparatus manufacturers – *LaFrance, Babcock* or *Gleason & Bailey* – who constructed their aerial ladder trucks *under license* from their patentees or inventors.

Most common by far was the *Hayes* – America's first mass-produced aerial ladder truck. The first half-dozen Hayes-patent aerial ladder trucks were built on the West Coast by the Kimball Mfg. Co. and the H. M. Black Co. Spurred by his success, Daniel D. Hayes then headed back east to seek out manufacturers to make and market his aerial ladder trucks there. In 1882, the *LaFrance Fire Engine Co.* of Elmira, New York, entered into an agreement with

The tillerman's seat and steering wheel were nestled *under* the aerial ladder in the distinctive Hayes design. Forward visibility must have been minimal at best. A substantial number of *LaFrance/Hayes* horse-drawn aerial ladder trucks were built for American fire departments from the 1880s into the early 1900s. This "second-size" 75-foot *LaFrance/Hayes* aerial was built for St. Louis, Missouri, sometime in the early 1900s. *MLC*

Mr. Hayes to manufacture and sell his aerial ladder trucks. By 1895, there were close to 200 Hayes aerial trucks in service across the U.S.

The Hayes ladder was raised by a long screw rod running above the turntable, which held a large nut with braces running to the butt strap of the ladder. On the Hayes truck, the tiller seat was *under* the main ladder – the tillerman had to peer around and through the stacked ground ladders before he swung his steering wheel to make a turn.

In 1884, Ernest F. Steck – inventor of the Steck monitor nozzle and Champion water tower and at the time superintendent of the Fire Extinguisher Mfg. Co. of Chicago, successor to the Babcock Mfg. Co., was issued a patent for an aerial ladder design under the well-known *Babcock* name. Steck's manually raised *Babcock* aerial utilized tall, vertically mounted worm screws on either side of the ladder to hoist the main ladder out of its bed. Crank handles and bevel gearing on the turntable did the rest. The tillerman on the Babcock was seated on *top* of the aerial ladder, vastly improving forward visibility compared to the rival Hayes design. The first Babcock aerial was delivered to New Haven, Connecticut, in 1897. Within two years, FEMCO had delivered two dozen more Babcock aerials to fire departments in the U.S.

Another aerial ladder design which found only limited success was the *Skinner*, invented by George Skinner of New York City. This ingenious ladder truck design featured a *telescopic* chassis frame. When the 90-foot, three-section ladder was raised, the rear axle and wheels were shifted forward, ostensibly to give the unit more stability. The first *Skinner Patent Hose Elevator* was built by the George Hannis Co. of Chicago and placed in service with the Chicago Fire Department in 1871 – the year of the Great Chicago Fire. Boston and St. Louis also purchased Skinner aerials.

In the 1880s, the E. B. Preston Company of Chicago built some aerial ladder trucks which incorporated – with minor variations – features of both the Hayes and Babcock designs. The Preston Company

Patented in 1884, the *Babcock Aerial* competed vigorously with the *LaFrance/Hayes* aerial in the late 1800s. Made by the *Fire Extinguisher Mfg. Co.* of Chicago, the Babcock aerial had distinctive forked butt and tall hoist stanchions on each side of the ladder. Along with the *Hayes,* American-LaFrance predecessor *International Fire Engine Co.* also built some Babcock-patent aerials in the early 1900s. *MLC*

is credited with delivering the first *all-steel* aerial ladder – a three-section, 90-footer – to St. Paul, Minn. in 1888, and another to Seattle shortly thereafter.

One of America's oldest fire apparatus manufacturers – the Gleason & Bailey Works of Seneca Falls, New York – introduced the *Dederick* aerial in 1895. The Dederick ladder was raised by cables attached to the top of the ladder. The fast-raising Gleason & Bailey-built *Dederick* aerial found much favor with the New York City (FDNY) Fire Department.

A peek inside a big-city firehouse in the early years of the new century would most likely reveal overhead harness ready to drop and a horse-drawn Hayes, Babcock or Dederick aerial waiting for the alarm bell to sound. At the fire, firemen furiously turning hand cranks raised the big ladder out of the bed, and, swung the turntable to effectively position the ladder in front of the burning building.

Indicative of the overwhelming popularity of the Hayes aerial, following is a list of the first aerial ladder trucks placed into service in America's largest cities:

New York, NY	1886 Hayes 85-foot
	1875 Scott-Uda *
Boston, MA	1873 Skinner
Baltimore, MD	1887 Hayes 65-foot
Philadelphia, PA	1881 Hayes 85-foot
Chicago, IL	1871 Skinner 84-foot
Detroit, MI	1883 Hayes
St. Louis, MO	1874 Skinner
Minneapolis, MN	1885 Hayes
Milwaukee, WI	1883 Hayes 1875 Buestrin*
Atlanta, GA	1886 Hayes 65-foot
San Francisco, CA	1872 Hayes
Los Angeles, CA	1881 Hayes

* Tested, but never officially placed in service

Incredibly, the Baltimore Fire Department used Hayes aerial ladder trucks *exclusively* until 1938. In the early 1900s, the B.F.D. repair shops retrofitted its substantial fleet of Hayes aerial ladders with Dahill

The *other* mainstream American aerial ladder truck design of the late 1800s was the *Dederick,* built by the Gleason & Bailey Works of Seneca Falls, New York. The quick-raising *Dederick* aerial was much favored by New York City Fire Department. The Detroit Fire Department motorized this late-1800s *Gleason & Bailey/Dederick* aerial with Ahrens-Fox Model "F" auto tractor in 1913. *MLC*

pneumatic hoists. These trucks were instantly identifiable by the huge cylindrical air reservoir tank on the turntable, at the base of the aerial ladder. Following the end of the horse-drawn era, these truly ancient aerials – with the tillerman still crouched in his seat under the main ladder – were hauled to fires by a succession of Mack, Seagrave and American-LaFrance motor tractors. After nearly three-quarters of a century of service, the last of Baltimore's venerable Hayes aerials wasn't hauled off to the scrapyard until 1962!

SPRINGS TO THE FORE

The next major advance in aerial ladder technology arrived in the early years of the new century. For some time, several fire apparatus manufacturers had been seeking ways to take the musclework – *manpower* – out of raising the heavy main ladder on horse-drawn aerial ladder trucks.

An early attempt in this area was the *Kaiser* compressed-air system. Developed in Europe, the *Kaiser Pneumatic* utilized a large air reservoir to lift the main ladder out of its bed. Charles A. Fox, superintendent of the American Fire Engine Co., and Edward J. Dahill, Chief of the New Bedford, Massachusetts, Fire Department, developed an improved pneumatic hoist based on the Kaiser system, which was retrofitted to a number of Hayes aerials on the Baltimore Fire Department. The Ahrens-Fox Fire Engine Company later employed this same basic hoist technology for its aerial ladder trucks in the 1920s.

Some water towers were *hydraulically* raised by attaching hose lines to a large tank on the truck's frame. Water pressure raised the tower mast to its single vertical operating position. Others were equipped with chemical tanks. When water was added to the soda solution, the resultant chemical reaction provided the lifting power.

The Seagrave Company of Columbus, Ohio was the first U.S. fire apparatus manufacturer to perfect a successful *spring-raised* aerial ladder hoist. Frederic Scott Seagrave, who had founded his business in De-

Seagrave introduced the industry's first *spring-raised* aerial ladder truck in 1902. The first two were delivered to New York City that same year. Taken circa 1904, this photo shows a Seagrave 85-foot aerial ladder truck on the rail loading dock at the company's plant in Columbus, Ohio. Note Seagrave's massive double-trussed aerial ladder design, and how the two long spring cylinders rose up with the ladder. *MLC*

troit in 1881 and moved it to Columbus, Ohio, ten years later, was a prolific inventor. In 1883, Mr. Seagrave was granted a patent for an extremely sturdy steel-reinforced truss-type ladder which found eager acceptance by fire departments all over the country. By the turn of the century, Seagrave & Co. was a major manufacturer of hand- and horse-drawn city service hook-and-ladder trucks and hose-and-chemical wagons.

Despite its well-deserved reputation as one of the nation's leading makers of fire department ladders, Seagrave did not enter the aerial ladder market until the early years of the new century. For some time, Mr. Seagrave had been seeking a way to eliminate the muscle power then needed to raise an aerial ladder. He found his solution in the captive energy of large compressed coil springs. The Seagrave Company delivered its first two aerial ladder trucks to New York City (FDNY) in 1902. Mr. Seagrave was granted a patent for his spring-raise hoist the following year. From this point on, improvements came rapidly.

The horse-drawn Seagrave aerial ladder truck had a distinctive double-trussed ladder, with prominent bowstring trusses above and below the main beam. The two-section wood ladder had a relatively high pivot point. The twin steel tubes in which the coil springs were enclosed rose off the turntable with the aerial ladder. The turntable was rotated 360 degrees by hand. The tillerman rode atop the main ladder. Seagrave aerials were made in 65-, 75- and 85-foot sizes.

In Elmira, New York, at about the same time, the *International Fire Engine Co.* was also working on a spring-raised aerial ladder. In early 1904, the former International Fire Engine Co. cartel was succeeded by a new corporate entity called the *American-LaFrance Fire Engine Company.* The American-LaFrance *Quick-Raising Aerial Ladder* was introduced that same year. In an interesting aside, young Addison Ward LaFrance – who would go on to form his own fire apparatus company – was one of the engineers assigned to this project. His father, Asa, and uncle, Truckson S. LaFrance, promised young Ward a new automobile if the spring-raised aerial was a success. It was, and A. Ward LaFrance collected his prize – a 1905 Rambler.

In 1908, a much-improved version of this aerial went into production with what was called the *American Automatic* hoist.

"The American Automatic Aerial Truck embodies several decided improvements over the (former) ALF aerial truck," an American-LaFrance ad in a 1909 issue of *Fire & Water Engineering* boasted. "No manual power required, no motors or air compressor necessary. The entire mechanism is self-contained. There are now in service 69 American-LaFrance aerial trucks. While it has been on the market but one year, 33 of them (American Automatics) have been sold within that time."

The *American Automatic* name was used on American-LaFrance's spring-raised aerial ladder trucks and water towers into the mid-1930s, until the introduction of the company's first hydraulic metal aerials.

As advanced as the automatic, spring-raised aerial ladder was, however, it was eclipsed by another even more spectacular technological advance that was al-

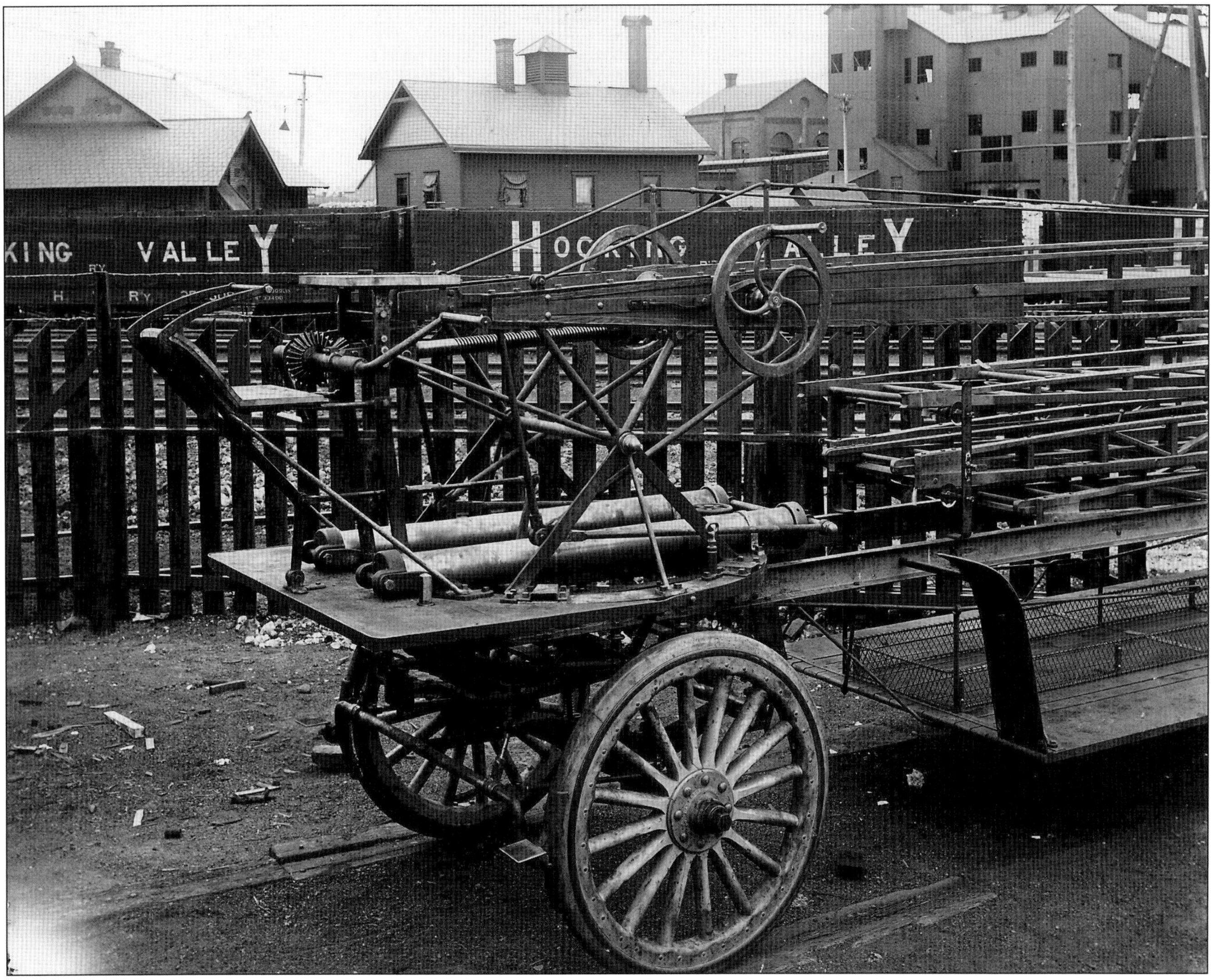

Hoist detail of an early Seagrave spring-raised aerial ladder truck, circa 1906. The heavy aerial ladder was raised out of its bed by strong steel coil springs housed in the two long tubes visible on the turntable. The upper or "fly" section was extended by cranking the large hand wheels on the sides of the ladder. The turntable was rotated manually. *MLC*

ready sweeping the nation. The horseless carriage had arrived. In 1905, Howe delivered a motor-driven pumper to the fire department in Larue, Ohio. The following year Knox and Waterous also delivered automobile pumpers to several fire departments. Seagrave introduced its first motor-driven fire apparatus in 1907. American-LaFrance was experimenting with self-propelled steam- and gasoline-powered vehicles, but would not introduce its first successful motor fire apparatus, the Type 5, until 1910.

The Seagrave Company exhibited America's first tractor-drawn, motor-driven aerial ladder truck at the 1908 convention of the International Association of Fire Engineers (IAFE) held in the company's home city of Columbus, Ohio, in October of that year. The 75-foot aerial ladder trailer was attached to a four-wheel tractor powered by a gargantuan 90-horsepower, four-cylinder Frayer-Miller air-cooled engine with 7-1/2-inch bore and 7-3/4-inch stroke. A contemporary commercial vehicle industry publication, *The Power Wagon,* noted that the chain-driven Seagrave was capable of traveling at a rate of 25 miles an hour over a five-percent grade – "three to four times the speed possible with a horse-drawn hook and ladder truck."

The truck's two-section, spring-raised wooden

Seagrave spring-raised aerials were made in three sizes — 65, 75 and 85 feet. This 75 footer was built for Washington, D.C., in 1907. Following introduction in 1902, Seagrave constantly refined and improved its hoist mechanism. Note the relocation of the coil spring lift cylinders to a new location *under* the base of the ladder, and Seagrave's distinctive double-truss design. Modern horse-drawn aerial trucks of early 1900s rolled on hard-rubber tires. *MLC*

aerial ladder was of a new non-trussed, single-beam design. Seagrave's – and the industry's – first motor-driven aerial ladder truck was delivered to Vancouver, British Columbia – which two years earlier had also purchased Seagrave's first three motor-driven fire engines – in early 1909.

As the first decade of the bold, new century drew to a close, modern rubber-tired Seagrave and American-LaFrance spring-raised, horse-drawn aerials were rapidly replacing now-outdated Hayes, Babcock and Dederick aerials on many metropolitan fire departments. Already past its infancy, the newfangled automobile was poised to put the legendary three-horse-hitch ladder truck and steam fire engine horse teams out to pasture.

This classic photo captures all the excitement of the fading horse-drawn era. Hooves flying and nostrils flared, beautifully matched three-horse hitch propels Truck Company No. 2's new Seagrave aerial ladder truck on a demonstration run in Columbus, Ohio, circa 1908. Crew of nine men certainly added to the weight. Note the two bowler-hatted dandies perched in the tree at the right! *MLC*

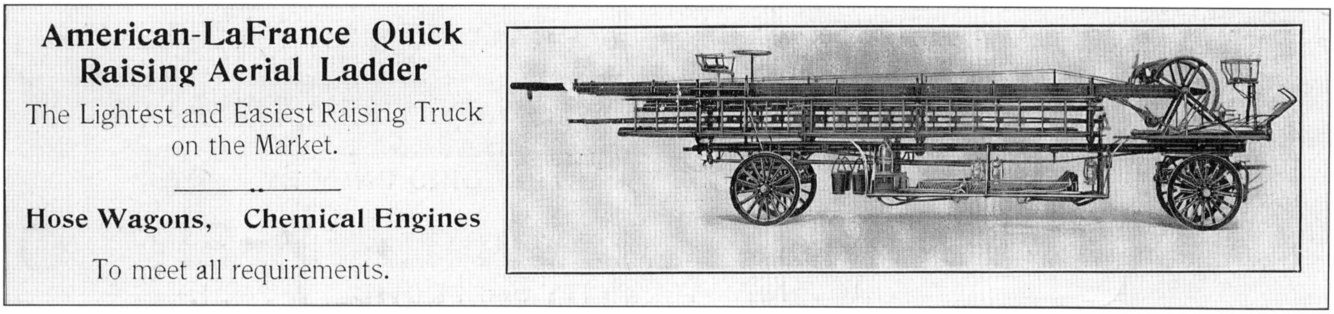

American-LaFrance introduced *American Automatic* spring-raised aerial ladder truck in 1903-04. Here's an advertising cut for the company's *Quick-Raising* spring-raised aerial ladder, circa 1910. *WMC*

American-LaFrance aerial ladder truck in action at tragic fire in downtown Cleveland on July 3,1908. Single-bank ladder bed has been stripped of ground ladders raised to second- and third-floor windows. Three-horse hitch remains hitched to the heavy truck. Three people were killed when fireworks were carelessly ignited in Kresge department store. *MLC*

Famed for its trademark trussed ladders, Seagrave emerged as the leader in aerial ladder truck design in early 1900s. In addition to the first successful spring-raised aerial, the Seagrave Company of Columbus, Ohio, also designed and built America's first *motor-driven* aerial ladder truck, which was exhibited at the 1908 annual convention of the International Association of Fire Engineers (IAFE) held in Columbus in August 1908. The company delivered its first motor-driven aerial — shown here — to Vancouver, British Columbia, early the following year. *WMC*

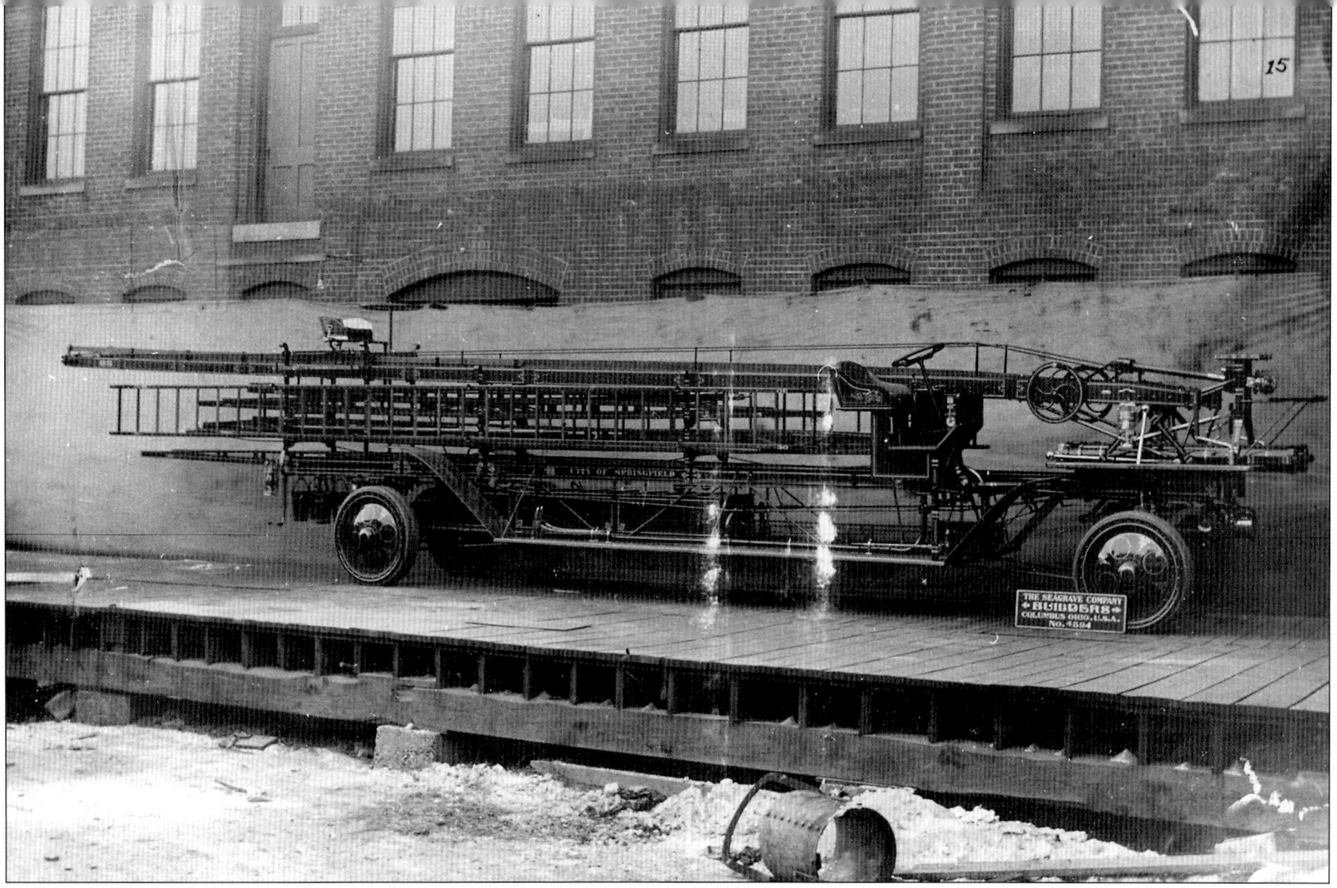

One of the first fire departments to motorize — starting in 1906 — Springfield, Massachusetts, purchased *three* of these battery-driven Seagrave 85-foot "sidesaddle" aerials, this one in 1910 and two more in 1912 and 1914. All three were built on four-wheel electric-drive chassis made by the *Couple Gear Freight Wheel Co.* of Grand Rapids, Michigan. Each wheel had its own electric drive motor. Note the driver's precarious perch on the right-hand side of the apparatus. *MLC*

CHAPTER TWO: THE TEENS-TRACTORIZATION

Motor-driven fire apparatus, which had sputtered uncertainly onto the scene in the early years of the 20th century, brought a stunningly swift end to the colorful horse-drawn era by the end of the new century's second decade. Almost overnight, it seemed, the romantic steam fire engine – which had been around for more than 60 years – was supplanted by the far more efficient, if less glamorous, automobile pumping engine.

In aerial apparatus, however, the most significant technological advance during the teens was in *motive power* rather than any major change or improvement in the aerial ladder itself. As the second decade of the century began, virtually all of the aerial ladder trucks in service in the United States and Canada were still drawn by horses. When the turbulent teens drew to a close in the years following the First World War, the vast majority of them were pulled by motor tractors.

Exactly which fire department was the first to place an auto tractor under a previously horse-drawn ladder truck is not precisely known, but Allentown, Pennsylvania, had to be among the first. In 1909, the Allentown Fire Department installed a locally built Mack four-wheel tractor under its Dahill air hoist-equipped Hayes aerial ladder truck. Thousands of similar conversions would soon follow all over the country. The conversion itself was remarkably simple. The front axle gear and wheels of the formerly horse-drawn ladder truck were removed, and what was now a *trailer* was attached to a "fifth-wheel" hitch at the rear of the auto tractor's frame.

No more horses to feed, hay bales to store, harnesses to repair or stables to clean! Best of all, the ladder truck could now keep up with the new motorized pumper and hose wagon.

The motorization of formerly horse-drawn aerial

During the teens, many fire departments converted their horse-drawn ladder trucks and water towers to motor power by replacing the horses with two- or four-wheel automobile tractors. Seagrave supplied this heavy-duty aerial truck tractor to Spokane, Washington, in 1911. Note the carbide head- and spot-lights and the long-handled crank for the gargantuan six-cylinder engine. *MLC*

apparatus was also driven by financial considerations. Many cities had significant taxpayer-funded investments in fleets of still-serviceable, not very old horse-drawn city service hook and ladder trucks, aerial trucks and water towers. It didn't make fiscal sense to discard a perfectly good horse-drawn apparatus with many years of service life still in it and replace it with an expensive, new factory-built job. It was much more practical and cost-effective to simply replace the horses with *horsepower*. So, as cities began to replace their steam fire engines and horse-drawn hose wagons with automobile apparatus in the teens, they similarly also began to motorize their ladder trucks.

A plethora of early motor truck manufacturers were more than willing to help. Two distinctly different types of motor tractors were available for this purpose. The simplest solution was a standard commercial truck – a four-wheel motor tractor built by companies such as White, Mack, GMC or FWD. The other option was a special-purpose, two-wheel, front-drive tractor designed to bolt to the front end of the formerly horse-drawn apparatus.

Interestingly, some of the earliest conversions were battery-powered vehicles or gas-electric hybrids. The Couple Gear Freight Wheel Company of Grand Rapids, Michigan, offered a variety of pure electric or gas-electric four-wheel tractors and propulsion components. New York City's first such conversion was a Webb-Couple Gear tractor purchased to motorize Water Tower No. 1's 1894 Hale in 1911. Philadelphia, Pennsylvania, purchased a number of battery-powered tractors built by the Commercial Truck Company of America of Newark, New Jersey.

But by far the most successful conversion tractor was the *Christie* made by the Front Drive Motor Company of Hoboken, New Jersey. A former racing car driver and prolific inventor John Walter Christie was also a tireless promoter of front-wheel-drive technology. In 1911, Christie constructed a prototype two-wheel tractor designed to replace the horses under formerly horse-drawn steam fire engines. Christie demonstrated his tractor to the chief officers of the New York City Fire Department (FDNY) – which fortuitously had just begun the motorization of the nation's largest metropolitan fire department. The chiefs were obviously impressed, immediately placing an order for 28 of the steamer tractors. The FDNY eventually purchased more than 300 Christie front-drive tractors.

The Christie front-drive tractor was built in two versions – one specifically for steamers and another somewhat different design for motorizing heavy aerial ladder trucks and water towers. On the far more

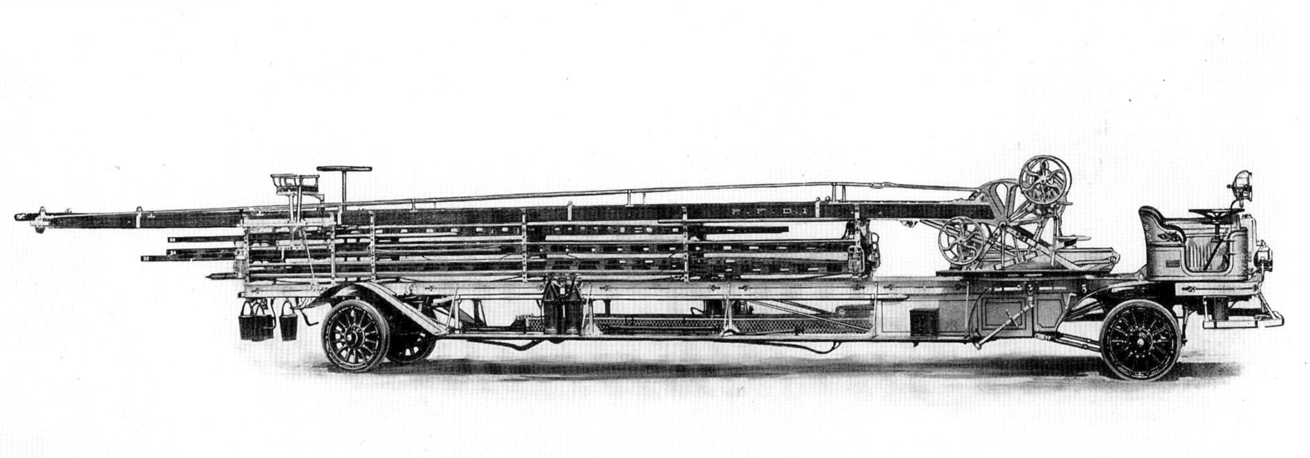

GASOLINE-ELECTRIC FOUR WHEEL DRIVE AND STEER AERIAL TRUCK
75 Horse Power—Four Cylinder Motor

American-LaFrance introduced a series of gasoline-powered and hybrid gas/electric straight-frame aerial ladder trucks during the early teens. Here are two of them. Above: *Type 16* gas/electric 85-foot aerial powered by 75-horsepower four-cylinder motor built for Portland, Oregon, in 1913. Below: gasoline-powered six-cylinder 85-foot *Type 18* aerial delivered to Jersey City, New Jersey, also in 1913. On both of these long trucks with tiller rear steering, the driver and officer were separated by the engine hood. *WMC*

common steamer version, the transversely mounted, four-cylinder engine was mounted under a low, beveled hood. On the aerial truck version, the motor was located under the driver's seat. The steering wheel and driver's low toeboard extended far out in front of the vehicle.

Strong visual characteristics of the extremely successful Christie tractor included huge cast steel disc wheels and a massive flywheel on the right-hand side of the apparatus. During the teens, Christie tractors were attached to just about every make of aerial then in service. Many Seagrave, American-LaFrance, Hayes, Babcock and other formerly horse-drawn aerials and water towers were repowered with Christie tractors, significantly extending their service lives. In 1918, the FDNY purchased five Combination Ladder Company straight-frame aerials with Christie tractors. Between 1912 and 1919, the Front Drive Motor Company delivered more than 600 Christie tractors to fire departments all over the United States. C. J. Cross, a Christie rival, also built a few front-drive aerial truck tractors.

Two other pioneer motor fire apparatus manufacturers – *Webb* and *Boyd* – built aerial ladder trucks of their own design in the early motor era. A famed racing driver, *Al C. Webb* converted a Buick roadster into a chemical car for the Joplin, Missouri, fire department in 1907. The following year his fledgling firm introduced a successful motor pumper on a Thomas Flyer chassis. Located at various times in Joplin, Vincennes and St. Louis, Missouri, the Webb Fire Apparatus Co. eventually moved east, to Allentown, Pennsylvania. By 1912 the company had added big aerial ladder trucks to its product line. Of the four-wheel, straight-frame type, the Webb aerial was powered by a midship-mounted engine which delivered electrical power to motors in all four wheels through the Coupe Gear hybrid drive system. A distinctive feature of the Webb aerial was a huge ¾-radius gear quadrant immediately behind the driver. The two-section 65-, 75- or 85-foot aerial was spring-raised.

James Boyd & Brother of Philadelphia also built more than a dozen aerial ladder trucks of the four-

Seagrave built fewer than a dozen of these odd "sidesaddle" straight-frame aerials between 1910 and 1919. While several were built with Couple Gear electric drive, this one delivered to Canton, Ohio, in 1913 had conventional gasoline power. The six-cylinder engine is under the louvered compartment visible just behind the front wheels. All had right-hand drive. *MLC*

wheel, straight-frame style – five of them for its home city of Philadelphia, with Couple Gear gasoline-hybrid powertrains. The Boyd aerials differed from competitors in that their 65-,75- or 85-foot two-section wood aerial ladders were not spring-raised. The Boyd aerial was raised by motor power through a power takeoff.

In 1916, the Ahrens-Fox Fire Engine Company of Cincinnati, Ohio, delivered a battery-powered aerial ladder truck to New Bedford, Massachusetts. The company's first aerial truck, the straight-frame rig with 75-foot, two-section Pirsch ladder was built on the Couple Gear gas-electric chassis with Dahill pneumatic aerial hoist – which, incidentally, was invented by Edward F. Dahill, Chief of the New Bedford Fire Department.

Regardless of the builder, the Couple-Gear electric and hybrid gasoline-electric aerials were notoriously slow. Batteries depleted, they sometimes had to be ignominiously towed back to their stations. By mid-decade, the bulky dual-mode hybrid technology had already been eclipsed by vastly more efficient and reliable gasoline engines.

The nation's two leading fire apparatus manufacturers – American-LaFrance and Seagrave – produced a remarkable variety of aerial ladder trucks through the teens. Perhaps the most unorthodox of these were the dozen or so "side-saddle" straight-frame aerials that Seagrave delivered between 1910 and 1919. The six-cylinder engine in the all-wheel-drive, all-wheel-steer Model "M" was located under the chassis frame near the front of the apparatus. The driver sat in a bucket seat bolted to the side of the rig about a third of the way down the right-hand side of the apparatus, beside the aerial ladder. How the driver *ever* saw what was ahead of him beyond the aerial truck's cluttered turntable remains a mystery. The model "M" was built in 65-, 75- and 85-foot sizes.

In the early to mid-teens, industry leader *American-LaFrance* introduced three long, straight-frame, four-wheel type aerial hook and ladder trucks of slightly varying design. Introduced in 1912, the *Type 16* was a four-wheel-drive-and-steer truck with hybrid propulsion system. A 75-horsepower, four-cylinder ALF Type 10 gasoline engine delivered power to General Electric railway-type generators mounted in all four wheels. With its large rectangular radiator, 65-, 75- or 85-foot two-section wood aerial ladder and American Automatic quick-raise spring hoist, just 16 *Type 16s* were built between 1912 and 1916.

The two-wheel, front-wheel-drive American-LaFrance *Type 18* aerial was of the same general configuration but had a six-cylinder gasoline engine under a long gabled hood between the two front seats. The New York City (FDNY) Fire Department purchased twenty-five 65- and 75-foot Type 18 aerials in 1913. The later front-drive *Type 25* aerial was also powered by a big six-cylinder engine nestled between the two front seats, with a taller radiator and cowcatcher-style wraparound front bumper.

In addition to conventional type four-wheel tractors, both of these companies introduced all-new, two-wheel front-drive tractors in the mid-teens. Introduced in 1915, American-LaFrance's ubiquitous

Albany, New York, converted its formerly horse-drawn 1907 American-LaFrance aerial to modern motor power with the addition of a 1913 *Alco* four-wheel tractor built by the American Locomotive Works. Note the huge gear quadrants at the base of the two-section wooden aerial ladder. *Albany Fire Department*

Type 31 became the company's most successful two-wheel, front-drive auto tractor with more than 300 sold before the last one was delivered in 1929. The Type 31's 75-horsepower four- or 105-horsepower six-cylinder engine was mounted ahead of the front axle. Power was transmitted directly to chain-drive jackshafts and sprockets behind the front wheels. The Type 31-4 or 31-6 could be ordered as either a bolt-on conversion unit to convert former horse-drawn aerials or water towers to gasoline power, or as a complete, straight-frame aerial or water tower. The versatile Type 31 was also available as a steam fire engine conversion tractor, with rocking trunnion between the tractor and heavy steam fire engine body. The Type 31 front-drive aerial was built in 55-, 65-, 75- and 85-foot sizes with ALF American Automatic hoist and two-section, spring-raised main ladder.

Members of the Cleveland Fire Department pose with the city's first motorized aerial ladder truck — a formerly horse-drawn 1908 American-LaFrance *American Automatic* spring-raised 85-foot aerial motorized with 1913 Seagrave AC90 air-cooled auto tractor. Detroit also motorized a late-model horse-drawn aerial with one of these massive Seagrave auto tractors. *WMC*

Seagrave's answer to ALF's Type 31 was the *Model "K"* – a heavy-duty four- or six-cylinder two-wheel front-drive tractor introduced at about the same time. The Model "K" tractor's engine was housed under a gable hood between the two front seats. Like the Type 31, the "K" was available as a bolt-on conversion unit or as a complete straight-frame 55-, 65-, 75- or 85-foot spring-raised tillered aerial. A steamer version of this heavy-duty tractor was also catalogued. Seagrave's Model "K" remained in production for about a decade with the last one delivered sometime in the mid-1920s.

Both companies also offered complete factory-built tractor-trailer type aerial ladder trucks. Most of these big-city types were powered by six-cylinder engines and had 75- or 85-foot spring-raised wood aerials and rolled on solid rubber tires with huge artillery-type wood spoke wheels on the trailer. The steering wheel was removed and the tillerman's seat swung away to one side when the aerial ladder was to be raised.

American-LaFrance's big tractor-drawn aerial was the *Type 17*, which denoted a four-wheel-type motor tractor attached to a 65-, 75- or 85-foot aerial trailer with tiller rear steering. The *Type 17* designation also applied to ALF's four-wheel tractor, which could be purchased separately for attachment to a formerly horse-drawn aerial ladder truck or water tower. Introduced in 1912, the *Type 17* was built into the mid-1920s. The vast majority of Type 17s were powered by ALF six-cylinder engines. Only half a dozen or so *Type 17-4s* were sold, mostly as conversion tractors for older city service hook and ladder trucks.

The majority of the aerial ladder trucks built in the teens were conversions of horse-drawn rigs. For every complete new tractor-drawn aerial built, many more older units were converted to gasoline power by the attachment of two- or four-wheel auto tractors. The most common of these heavy commercial-type four-wheel motor tractors were Whites, Macks and FWDs. Robinson and American & British also delivered some two-wheel conversion-type ladder truck tractors.

During the teens and early 1920s, the Chicago Fire Department motorized its large fleet of city service hook and ladder trucks and aerial trucks with heavy-duty motor tractors, primarily Mack AC *Bulldogs,* and Whites. The C.F.D. purchased its first factory-built, tractor-drawn aerials – five 85-foot Seagraves – in 1928.

Except for a relatively few with Dahill-type pneumatic or engine-driven hoists, most of the aerial ladder trucks built in the teens were of the two-section, wood spring-raised type. Why this overwhelming reliance on coil springs for ladder lifting power? Here's how The Seagrave Co. justified their use in a sales catalog of the era:

"*Because they are more dependable than any other device. They will operate for years without serious loss of power, and means are provided for adjusting them whenever that becomes necessary. Turning of the three adjusting screws in the forward end of each spring tube compensates for loss of spring tension. The use of springs entails less mechanism than the use of liquids or air, and springs are not affected by climatic conditions.*"

"*The hoisting power is derived from heavy coil springs in tubes mounted on the turntable. When these springs are released by operating a small lever, they automatically raise the main ladder from its bed to perpendicular in about five seconds. A friction brake easily operated by one hand is provided for controlling the speed or stopping the ladder in any desired position while it is rising to the perpendicular. When the ladder is stopped it may be easily locked in that position: when released it will continue to rise to the perpendicular. A large screw and split nut are used for lowering the ladder or controlling it when necessary.*"

Seagrave's Canadian cousin — *W.E. Seagrave Fire Apparatus* of Walkerville, Ontario — delivered this imposing motor-driven 85-foot aerial ladder truck to Montreal, Quebec, in 1914. This would be the final year for the double-trussed style aerial ladder, which was replaced by a lighter, more conventional solid single-beam design. *WMC*

The top, or fly section of the ladder, was extended and the turntable rotated by cranking large handwheels.

Los Angeles, California, was a large purchaser of early Seagrave motor fire apparatus. The L.A.F.D. took delivery of this 75-foot Seagrave aerial ladder truck with new-style single-beam ladder in 1914. Los Angeles apparatus of this era was painted white with maroon wheels. *MLC*

Despite rapid transition to motor-driven fire apparatus, some fire departments continued to order horse-drawn equipment well into the teens. Seagrave delivered this horse-drawn third-size 65-foot spring-raised aerial ladder truck to Charlottesville, Virginia, in 1915. Note the dual chemical tanks and chemical hose basket just ahead of the rear wheels. *MLC*

American Automatic fast-raising spring-hoist aerial ladder trucks made by American-LaFrance were available in four sizes — 55, 65, 75 and 85 feet. Relatively few 55-foot aerials were sold. Here's one of them — a late-model horse-drawn *American Automatic* 55 footer built for the *Senate Hook & Ladder Co. No. 1* of Ossining, New York, circa 1915. Note the rubber tires and brake drums on front wheels. *MLC*

Former racing driver Al C. Webb began building motor fire apparatus in Joplin, Missouri, in 1907. Later located in Vincennes and St. Louis, Missouri, *Webb Motor Fire Apparatus Co.* finally relocated to Allentown, Pennsylvania, where the company began making big aerial ladder trucks — most with hybrid *Couple Gear* gas-electric drive. This Webb aerial ad was published in a fire service trade journal in 1912. *WMC*

Another pioneer builder of motor aerial ladder trucks was *James Boyd & Bro.* of Philadelphia. Boyd built more than a dozen straight-frame motor aerial ladder trucks of this style between 1913 and 1916. This Boyd 85-foot aerial was delivered to Akron, Ohio, in 1914. Note passenger-car-style radiator and engine hood visible just behind the turntable. *Tim W. Elder*

Like other big cities, Boston Fire Department had considerable investment in still-serviceable horse-drawn fire apparatus. Conversion to motor power began in earnest in the teens. Typical example is Ladder 14's formerly horse-drawn American-LaFrance aerial, which was converted to motor power by the attachment of a 1916 Robinson six-cylinder, two-wheel auto tractor. *WMC*

In most cases, powerful motor-driven fire apparatus could go where horses couldn't in deep snow. Here, a 1913 American-LaFrance *Type 16* aerial for Louisville, Kentucky, plows through drifts during winter road test prior to delivery. Note wrap-around front bumper and separate chariot-style seats for driver and officer flanking large rectangular radiator. *MLC*

One of the most successful two-wheel conversion tractors was the *Christie,* built by the *Front-Drive Motor Co.* of Hoboken, New Jersey, between 1911 and 1919. New York City purchased more than 300 Christie front-drive tractors to convert formerly horse-drawn steam fire engines and ladder trucks to gasoline power. Woonsocket, R.I., bolted this Christie Front-Drive tractor to a formerly horse-drawn Seagrave 75-foot aerial circa 1915. *MLC*

Yet another front-drive aerial design by American-LaFrance was the six-cylinder, gasoline-powered *Type 25.* In separate bucket seats, driver and officer sat well back behind long hood and tall radiator. This 1915 *Type 25-6* aerial with 75-foot ladder was delivered to Fall River, Massachusetts. *WMC*

Introduced in 1915, American-LaFrance's most successful two-wheel type front-drive tractor was the pug-nosed *Type 31.* Available in four- and six-cylinder versions, more than 300 were built before Type 31 production ended in 1929. During the teens, many Type 31s were sold as conversion tractors, but this 85-foot *Type 31-6* aerial was built as a complete unit for Jersey City, New Jersey, in 1916. *WMC*

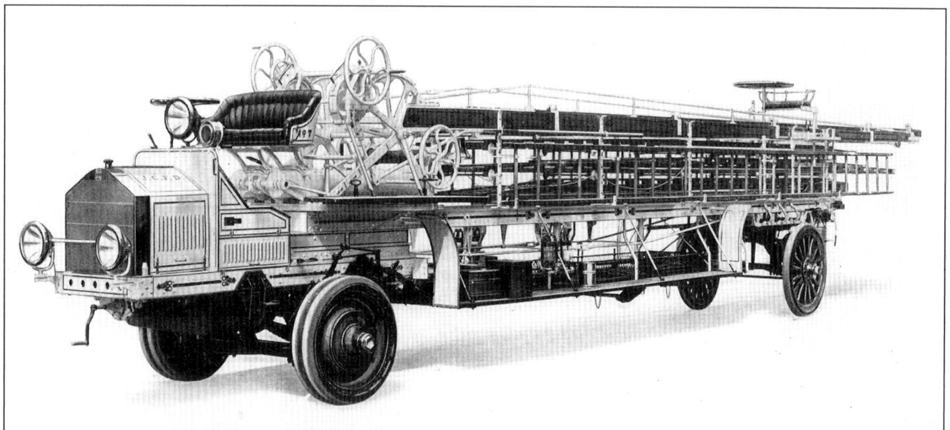

TYPE 31—GASOLINE TWO WHEEL FRONT AERIAL HOOK AND LADDER TRUCK
Six Cylinder Motor—105 Horse Power

Although the *Hayes* aerial faded away following the introduction of more modern spring-raised aerials early in the new century, some fire departments kept their venerable Hayes ladders in service for many more years. The Baltimore Fire Department, in fact, used nothing *but* Hayes aerials into the 1930s, regularly refurbishing and modernizing these old "sticks" — all with *Dahill* air hoists. The Fire Museum of Maryland in Baltimore owns this ex-B.F.D. 1908 Hayes 85-foot aerial with 1918 Mack AC "Bulldog" tractor. *Fire Museum of Maryland*

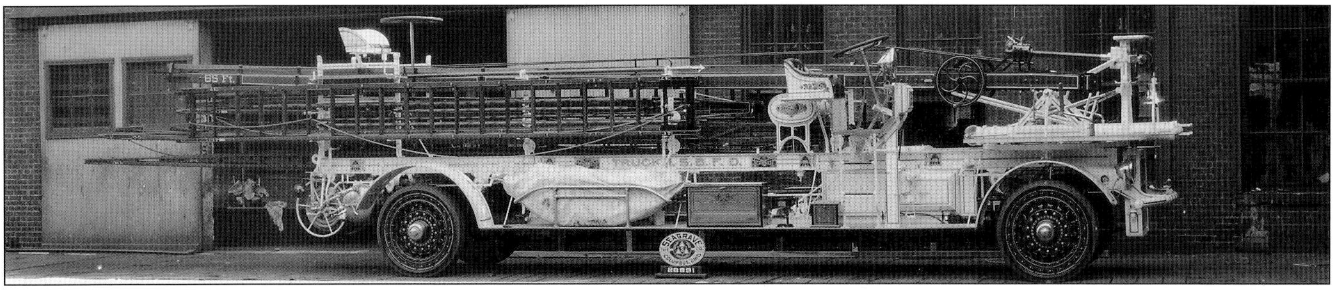

The last of the dozen "sidesaddle" straight-frame aerials built by Seagrave in the teens went to Santa Barbara, California, in 1919. Note the large rectangular radiator below the front of the turntable, quarter-fold life net on the running board, and chemical tank and hose reel under the rear of the frame. *MLC*

1916 American-LaFrance *Type 31* front-drive aerial at Great Lakes Chapter SPAAMFAA Muster in Frankenmuth, Michigan. *WMC*

American-LaFrance built just *one* Type 31 front-drive aerial with 55-foot aerial ladder. Register #1871 went to Harper's Ferry, Ohio, in 1918. Four-cylinder *Type 31-4* had extremely short hood. Note vertical steering column and how the driver's seat is attached via curved arm to the front end of the ladder turntable. *MLC*

By the end of the decade, industry leader *American-LaFrance* was building two distinct styles of aerial ladder trucks — the straight-frame, front-drive *Type 31* and the articulated, tractor-drawn *Type 17* shown here. The Type 17 could be ordered as a four-wheel conversion tractor or as a complete tractor-trailer aerial like this one built for Rocky Mount, North Carolina, in 1919. *MLC*

Peter Pirsch & Sons Co. of Kenosha, Wisconsin, built only a few aerial ladder trucks prior to 1930. This early *midship-mount* aerial on locally made Winther chassis was delivered to Virginia, Minnesota, in the early 1920s. Note the large hydraulic lift cylinder behind driver's seat. The sunken tiller seat and steering wheel are barely visible. *MLC*

CHAPTER THREE: THE 1920s

"The period of experiment and development in the motor car field has passed, and radical changes in design are not to be expected," the foreword in an American-LaFrance fire apparatus catalog published in the early 1920s noted. *"In the application of the motor vehicle to Fire Department service, we have not attempted to develop anything essentially new in principle, or radically different from well-tried and accepted standards."*

This status-quo commentary neatly summarizes progress in aerial ladder truck development through the fast-paced, increasingly prosperous decade historians have dubbed the *Roaring Twenties.* While there were indeed continuing refinements in aerial ladder truck design, there were no major breakthroughs or advances.

As the third decade of the Twentieth Century unfolded, there were just two manufacturers of aerial ladder trucks in America – the *American-LaFrance Fire Engine Company* of Elmira, New York, and *The Seagrave Company* of Columbus, Ohio. All three of the nation's pioneer makers of aerial ladders – *Hayes, Babcock* and *Gleason & Bailey (Dederick)* – had long since faded away with the bygone horse-drawn era.

But two other prominent U.S. motor fire apparatus manufacturers – *Mack* and *Ahrens-Fox* – charged into the increasingly competitive aerial ladder truck market in the 1920s.

Commercial truck maker Mack (or, more accurately, the International Motor Company of Allentown, Pennsylvania) had joined the growing number of U.S. fire apparatus builders with the introduction of its slope-nosed Model AC – better known as the *Bulldog* – in 1916. Mack introduced its first pumper three years later and began building tractor-drawn aerial hook and ladder trucks with brawny AC tractors in the early 1920s. New York City purchased 21 Mack AC-7 tractor-drawn 65- and 75-foot tillered aerials between 1921 and 1923 – and a single AC-6 four-wheel tractor to haul one of its formerly horse-drawn water towers. The Mack aerials had two-section spring-raised wood aerial ladders and oversized wood-spoked artillery wheels shod with hard-rubber tires on their tillered trailers.

Founded in 1910, the *Ahrens-Fox Fire Engine Company* of Cincinnati, Ohio, was famed for its distinctive front-mounted piston pumpers – "the fire engine with the big silver ball up front." This iconic company had delivered a straight-frame hybrid gas-electric aerial ladder truck to New Bedford, Massachusetts, in 1916 – the sole aerial apparatus this company had constructed up to now. But in 1923, Ahrens-Fox entered the ladder truck segment of the thriving and still-growing motor fire apparatus industry in spectacular style. In addition to a full line of four- and six-cylinder city service hook and ladder trucks, Ahrens-Fox simultaneously introduced a companion series of big tractor-drawn tillered aerial ladder trucks.

Drawn by six-cylinder, four-wheel tractors with Ahrens-Fox's classic gabled hood and radiator, the company's first production aerials were equipped with a much-improved Dahill pneumatic hoist. A

Two *American-LaFrance Type 17* tractor-drawn aerials in action at a commercial building fire in an unknown city in the 1920s. Note the two hose lines running up the aerial ladder to the building's roof. Looks like they may be getting ready to raise the second aerial in the foreground. *MLC*

Mack Trucks began building spring-raised aerial ladder trucks in the early 1920s — all of them hauled by the company's brawny *Model AC* "Bulldog" tractor. New York City purchased 21 of these Mack 65- and 75-foot aerials in 1921 and 1923. The Mack *Model AC-7* tractor-drawn aerial in this photo was built for the Harry Howard Hook & Ladder Co. No. 1 in Port Chester, New York, in the mid-1920s. *MLC*

pair of transverse-mounted air tanks at the front of the trailer frame supplied compressed air to lifting cylinders on each side of the aerial ladder's base on the turntable. One man could easily raise and extend the two-section wood aerial ladder, which was available in 65-, 75- and 85-foot sizes.

Ahrens-Fox also pioneered a new method of carrying the portable ground ladders on its ladder trucks, which would be quickly emulated by the rest of the industry. Up to now, ground ladders ranging in length from 12 to 55 feet were carried in a single stack, with additional long ladders hung on the sides of the single-bank ladder bed. Ahrens-Fox fathered the *double bank* configuration, with two side-by-side stacks of ladders. In addition to increasing ladder-carrying capacity, Ahrens-Fox's breakthrough design also reduced the vehicle's center of gravity, enhancing safety for the driver and crew.

In the mid-1920s, a famed German fire apparatus manufacturer – *Magirus* – introduced a radically different type of aerial ladder truck to America. But even though Magirus's European style *rear-mounted* aerial ladder truck design was four decades ahead of its time, it found only limited initial success with the somewhat conservative, tradition-bound North American fire service. The mechanically raised Magirus aerial's turntable was mounted at the *rear* of the apparatus, with the nested ladder facing forward over the cab. The typical American aerial had a midship-mounted turntable with the long aerial ladder overhanging the rear of the rig. These very long trucks required a tillerman to steer the rear wheels.

In August 1924 the Magirus Fire Appliance Co. of Ulm, Germany, demonstrated one of its *Combination Aerial Ladder and Water Tower* trucks at the annual convention of the International Association of Fire Engineers (IAFE) in Buffalo, New York. In that impressive demonstration, the 90-foot Magirus ladder was raised and extended to its full height in an astounding 23 seconds! At just 27 feet, the compact Magirus aerial ladder truck was half the length of a conventional American aerial truck of the day and only slightly longer than a pumper. The metal-braced three- or four-section Magirus ladder could be raised, extended and rotated by one man, and no tillerman was needed to steer the rear end of the apparatus.

While the Magirus's short wheelbase allowed it to squeeze into places no conventional American aerial could reach, there was one major drawback. The continental prototype aerial carried few, if any, ground ladders and had to be accompanied on alarms by another ladder truck.

Encouraged by the positive response to this demonstration, the German firm immediately set up a subsidiary company to market the Magirus aerial ladder truck in America. The first U.S. delivery went to East St. Louis, Illinois. Between 1925 and 1932, the American Magirus company delivered seven of these stubby rearmount aerial trucks to fire departments in the U.S. – to East St. Louis, Illinois, Aberdeen, South Dakota, Gary, Indiana, Milwaukee, Wisconsin, and Sidney, Lima and Galion, Ohio. Four more were purchased by Canadian fire departments – Calgary and Edmonton, Alberta, and two for Montreal, Quebec, bringing total North American sales to eleven. The deepening Depression crippled sales in the 1930s. Two decades would pass before the Magirus rearmount reappeared in the U.S. in the early 1950s. But widespread North American acceptance of the rearmount aerial configuration – virtually universal everywhere else in the world *except in* America – would not begin until the late 1960s.

All the way up! American-LaFrance *Type 17-6* tractor-drawn *American Automatic* 75-foot spring-raised aerial undergoes test on side street near the Elmira, New York, plant. Note the ladder pipe attached to the underside of the bed section of the ladder, permitting it to be used as a water tower. Note also the half-fold life net on the running board. *MLC*

Detroit Fire Department's Ladder 29 rolls out of quarters in October 1949. Their rig is a 1921 Seagrave 85 footer. Note the canvas windbreaker, Roto-Ray warning light, and coiled roll of three-inch hose for the ladder pipe. Original hard-rubber tires have been replaced with pneumatics. Seagrave used this gable hood design from 1914 into the mid-1920s. *MLC*

Two other well-known U.S. fire apparatus builders dabbled in aerial ladder trucks in the 1920s. The *Peter Pirsch & Sons Co.* of Kenosha, Wisconsin, delivered a new *type* of aerial truck to Virginia, Minnesota, in the early twenties. Mounted on heavy-duty Winther commercial truck chassis, the Pirsch aerial had a midship-mounted two-section main ladder – a progenitor of what would later come to be known as the *service aerial* – a conventional city service hook and ladder truck equipped with a short aerial ladder. And in 1925, the *Stutz Fire Engine Company* of Indianapolis, Indiana, built a tractor-drawn 85-foot aerial – the only aerial truck this short-lived company ever made – to Kansas City, Missouri.

Industry leaders American-LaFrance and Seagrave continued to refine their standard aerial ladder trucks. The tractor-drawn American-LaFrance *Type 17* and front-wheel-drive, straight-frame *Type 31* both utilized the American Automatic quick-raise spring hoist. By the mid-twenties, Seagrave had ceased production of its now-outdated "K"-Series front-drive service aerials. In 1926, American-LaFrance redesigned its standard motor fire apparatus, including the Type 17. The new *Type 117* aerial ladder and water tower tractor got a restyled front end with tall radiator and double-louvered hood and recessed cast metal cowl.

Mack Trucks introduced an uncommonly handsome new tractor-trailer aerial truck in 1928. Drawn by the company's new "B" Series tractor, the low-slung *Type 90* aerial truck boasted a powered *engine-driven* aerial ladder hoist.

Although virtually no one at the time knew the nation was on the brink of economic disaster, industry giant American-LaFrance announced an entirely new generation of pumpers and ladder trucks at the 1929 International Association of Fire Chiefs' convention in Birmingham, Alabama. The company's stylish new *Master Series* line incorporated such advances as left-hand steering, four-wheel brakes, a new heat-treated "fish belly" chassis frame for ladder trucks, hollow-spoke cast steel wheels with pneumatic tires and a fully-automatic closed cooling system. The pleasing new look included a long, low tapered hood, broad-shouldered radiator and full-crowned fenders. The milestone Master Series apparatus projected a powerful, integrated appearance with commanding "road presence" – the industry's first fire apparatus in which *appearance* was an important design consideration.

The company's new Master Series Type 217 tractor-drawn aerial boasted the longest one-piece offset chassis frame ever made and wide, low-set running boards. The totally redesigned Type 217 – available in 75- and 85-foot sizes – featured a simplified, uncluttered turntable and control pedestal with one-man operation. Introduced on the eve of the Depression, the 1929 Master Series line also included an all-new front-drive, straight-frame aerial truck. The new *Type 231* replaced the venerable *Type 31* which had been in continuous production since 1915. The new *Type 231* had a towering, flat-faced pug-nosed front end that incorporated the new Master Series radiator. All mechanical components were neatly enclosed behind the front-end sheet metal with handy cast aluminum step plates for ascent to the lofty cab. The new aerial was powered by an improved, slightly more powerful version of American-LaFrance's big six-cylinder engine and had a refined and improved *American Automatic* aerial ladder spring hoist.

Just as the Elmira, New York, plant was launching this state-of-the-art series in the final quarter of 1929, the stock market crashed and the nation began its long slide into a debilitating depression. The good times were *over.*

American-LaFrance built more than a hundred of these classic front-drive *Type 31 American Automatic* aerial ladder trucks between 1916 and 1929. Most were six-cylinder *Type 31-6*s like this one delivered to Oshkosh, Wisconsin, in 1923. Most had 75- or 85-foot ladders, but 65- and 55-foot sizes were also available. Note "cowcatcher" front bumper, disc wheels and tiller rear steering. *WMC*

Proud crew of Schenectady, New York's Truck Co. No. 1 pose with their new 1923 American-LaFrance *Type 31-6* front-drive 85-foot aerial in front of Central Fire Station on State St. Note cycle-type front fenders, Vesta drum headlights and curved gooseneck frame attaching driver's seat to the turntable. The seat rotated with turntable. *MLC*

Spring-Raise Aerial Operating Instructions

Despite their somewhat intimidating appearance and size, spring-raised wood aerial ladders were relatively simple to operate. Following are instructions for positioning – "spotting" – and operating a Seagrave 75- or 85-foot wood aerial of the late 1920s:

1. Spot the truck in position making sure there are no wires or other obstructions in the way.

2. Place chock blocks under the tractor's wheels.

3. Adjust jacks to axle housing.

4. Remove strap from ladder pipe (if it is to be used); swing the tiller seat and windshield out of the way, remove steering wheel and column.

5. Raise the brake lever to release the aerial ladder. Allow the ladder to rise to the height it is to be used. Push the brake lever *down* to stop the ladder in lock position. *Be sure the lever is in the full-down position.*

6. Step on the turntable lock to release the turntable. Loosen lock on the turntable rotation wheel, then turn to the position required.

7. Raise the upper (fly) section of the ladder to the desired height and make sure the locks are locked on the rungs and that the safety lock is on.

8. Put in handles and turn *clutch handle* to the right for making special adjustment to height – either lower or higher. When at the proper height, make sure the clutch handle is to the *left* before climbing the ladder. This operation is done by standing on the turntable and facing the tiller end.

9. To *lower* the aerial ladder, raise the ladder with cranks so that it is clear of the building. Lower the fly ladder coiling cable neatly on drum (watch your fingers!). Loosen the handle lock on the turntable then turn until the ladder is over the ladder bed. The step on lock on the turntable should click *locked* as the ladder is lined up with the truck.

10. To lower the aerial, put the clutch handle to the right after cranks are in place and crank down.

11. Have men in the rear to guide the ladder into position.

12. Turn the aerial cranks tight to put pressure on the aerial lift springs.

13. Put clutch handle to the left.

14. Release jacks and chock blocks.

Famed for its front-mount piston pumpers (the ones with the big silver ball out front), the *Ahrens-Fox Fire Engine Co.* of Cincinnati, Ohio, began building tractor-drawn aerial ladder trucks in 1923. Instead of the spring-raised hoists favored by rivals American-LaFrance and Seagrave, Ahrens-Fox aerials used *Dahill Pneumatic* hoists. Here, operator checks air pressure in dual air tanks at front of trailer frame. Note nickeled lifting cylinders on turntable. *MLC*

Factory-fresh 1923 *Ahrens-Fox* tractor-drawn aerial for Nashua, New Hampshire, undergoes test on street shortly after delivery. Note metal-tipped fly section and guide rope tied to fly section. Tillerman's seat pivoted out of the way when the two-section wood aerial ladder was raised. *MLC*

Classic builder's delivery photograph of Ahrens-Fox 85-foot tractor-drawn aerial delivered to Nashua, New Hampshire, in August 1923. Photo taken on campus of the University of Cincinnati. *MLC*

American-LaFrance *Type 17-6* tractor-drawn 75-foot aerial of early 1920s vintage, city unknown. Note half-fold life net and Sewell Solid Cushion hard-rubber tires. *Dan G. Martin*

Seagrave's competitor to rival American-LaFrance's *Type 31* two-wheel, front-drive tractor was the *Model "K"* introduced circa 1916 and built through the mid-1920s. Six-cylinder engine was located between seats. Like *Type 31, Seagrave "K"* was made in four- and six-cylinder versions designed to bolt onto former horse-drawn steamers and ladder trucks. Winnipeg, Manitoba, received a pair of Seagrave *Model K-756* 75-foot aerials in 1925. *WMC*

American-LaFrance factory photo of circa 1925 *Type 31-6 American Automatic Aerial* with two-section, 75-foot spring-raised aerial raised to full-vertical position but with upper (fly) section unextended. Note the skinny hard-rubber tires front and rear. *MLC*

Like other big cities, New York purchased some commercial-type four-wheel replacement tractors for older aerial ladder trucks and water towers. FDNY received 20 of these FWD tractors in 1926, including this one attached to Hook & Ladder Co. 27's Seagrave aerial. FDNY aerial specs required huge artillery-type trailer wheels. *MLC*

Mack became serious player in U.S. fire apparatus industry with introduction of its first six-cylinder engine in late 1920s. Kearny, New Jersey, took delivery of this impressive Mack 85-foot aerial pulled by 1928 *Model AP* tractor. Note pneumatic tires on front axle only. *MLC*

Despite their great length, tractor-drawn aerials were surprisingly nimble when handled by well-trained driver and tillerman. Here, Hempstead, Long Island's 1927 Ahrens-Fox 85-foot aerial swings around corner in local parade. Note hard-rubber tires on rear axle of the tractor. *Photo Courtesy Of Herb Barber*

Seagrave tractor-drawn 85-foot aerial truck of late 1920s, this one shipped to Rye, New York, in 1929. Company routinely photographed new apparatus on rail loading dock before loading into open-end boxcar for delivery to purchaser. Note full complement of Seagrave trussed ground ladders and gas-mask boxes on running board. *MLC*

In 1929, Mack introduced a totally new, advanced aerial ladder truck design with industry-exclusive *engine-driven* aerial hoist. Power takeoff from 150-horsepower, six-cylinder engine raised and lowered the two-section wood aerial ladder. Among first delivered was this 1929 *Type 19* tractor-drawn aerial with 85-foot ladder for Ardmore, Pennsylvania. *MLC*

In 1928, Ahrens-Fox extensively redesigned its custom chassis. The new look included softly rounded hood, cowl and radiator, new crowned front fenders and a heavy-duty double-bar front bumper. Paterson, New Jersey, received this 1929 Ahrens-Fox 75-foot, tractor-drawn aerial, which was photographed by the late Dick Adelman in 1950. *JFS/WMC*

American-LaFrance announced all-new *Master Series* fire apparatus in late 1929. Handsome new design included pumpers, city service ladder trucks and quads, tractor-drawn aerials and water towers, and straight-frame service aerials. New *Master Series* aerial featured redesigned turntable and improved *American Automatic* spring hoist. Among first completed was this 1929 *Type 217* tractor-drawn 75-foot aerial with hard-rubber tires on all six wheels. *WMC*

Clean, relatively uncluttered turntable of a Seagrave spring-raised aerial ladder of the late 1920s. Tension of coil springs was adjusted by turning protruding bolts on ends of the long spring tubes. Cranks on pedestal-rotated turntable and hand wheels on sides of extended upper "fly" section. *MLC*

American-LaFrance aerial ladder truck turntable, early 1930s. Firefighter in rubber coat and leather helmet has his right foot on spring-release lever, right hand on brake handle used to control ladder's angle. Note single-bank ladder bed, non-slip tread on turntable and step/cover over forward end of the spring cylinders. *WMC*

CHAPTER FOUR: THE 1930s- HYDRAULICS AND STEEL

In much the same way *coil springs* revolutionized aerial ladder truck design and operation in the early years of the new century, the development and introduction of *hydraulically operated, all-steel aerial ladders* in the1930s marked the next major technological advance in aerial firefighting in America.

More than anything else, however, the 1930s was overshadowed by the worst economic disaster in the nation's history. Within weeks of the stock market crash of October 1929, the entire country was plunged into a devastating depression. Fortunes were lost, factories closed and hundreds of thousands were thrown out of work. Production and sales of new cars and trucks dwindled to a fraction of what they had been in 1929. Dozens of auto and truck manufacturers – Pierce-Arrow, Auburn and Duesenberg among them – went out of business.

American fire apparatus manufacturers were also hard-hit by the depression. Cash-strapped cities and towns could no longer afford to buy new fire equipment. Orders for new pumpers and ladder trucks were few and far between, and production slowed to a trickle. Some plants operated only a few days a week and eagerly solicited repair and rebuilding work just to keep the plants operating.

Despite the grinding depression, however, the 1930s was also an era of astonishing progress and engineering advancement in the beleaguered auto industry. Even as the depression hit bottom in the early 1930s, automakers introduced bold, streamlined body styling and Olympian V-12 and V-16 engines. It was the golden era of the automobile.

Ahrens-Fox pioneered a combined aerial ladder/water tower with *Tower Aerial* patented in 1933. Steel upper- and lower-section handrails permitted simultaneous use of *two* ladder pipes, one on each section. First three Ahrens-Fox *Tower Aerials* were delivered to Newark, New Jersey, in 1930. Two Cincinnati Fire Department Ahrens-Fox piston pumpers pump into one of the Newark aerials in pre-delivery test. *MLC*

Industry leader American-LaFrance – which had introduced its all-new *Master Series* fire apparatus in the latter half of 1929 – took the rest of the industry by surprise with its announcement of a V-12 engine in mid-1931. With a narrow 30-degree bank angle, the 754-cubic inch, 240-horsepower "312-A" V-12 was essentially two six-cylinder engines that shared a common crankshaft. Following a transcontinental demonstration tour, the company delivered its first V-12-powered pumper to Winfield, Kansas, in August of that year. The big V-12 proved ideal for hauling heavy aerial ladder trucks and water towers. The first aerial truck powered by this landmark fire service engine was delivered to New Rochelle, New York, later that same month.

In 1929, the Ahrens-Fox Fire Engine Co. extensively redesigned its aerial ladder, adding steel-truss handrails to the second (fly) section of the wooden aerial to permit the ladder's use as a water tower. A ladder pipe fed by three 2-1/2-inch hose lines was attached to the tip of the fly section. During the early 1930s, Ahrens-Fox delivered a number of these tractor-drawn "*Tower Aerials*" – the beginning of the end for the single-purpose big-city water tower.

Although well known for its sturdy wooden ground ladders, the Peter Pirsch & Sons Co. of Kenosha, Wisconsin, had built only one or two aerial ladder trucks prior to 1930. But in the 1930s, Pirsch aggressively entered the highly specialized, increasingly competitive aerial ladder truck business.

Almost since the turn of the century, nearly all of the aerial ladders made in North America were of the two-section, manually extended type. While the main ladder was raised from its bed by releasing a pair of heavy coil springs, the upper, or fly section, was extended and the ladder turntable rotated by firefighters furiously turning big handwheels. In 1931, Peter Pirsch & Sons introduced the industry's first *hydro-mechanical* aerial ladder hoist. One man could easily raise, extend and rotate the two-section, 85-foot wood aerial ladder from a control pedestal on the truck's turntable.

Drawn by a heavy-duty White tractor, the first of these pioneering aerial ladder trucks was delivered to Spokane, Washington. Pirsch subsequently delivered a number of custom-chassis hydro-mechanical tractor-drawn 85-foot wood aerials to Minneapolis, Memphis, and Washington, D.C.

At about the same time, Pirsch introduced yet another game-changer. Designed for easy installation on any standard city service hook and ladder truck, the *Pirsch Junior Aerial* was a simple, manually operated two-section 50-foot wood aerial ladder which closely resembled those commonly used by public utility companies. Later versions were equipped with an electric hoist. The Pirsch Junior marked the advent of what came to be known as the *Service Aerial* – a city service ladder truck equipped with a junior-sized (50- to 55-foot) aerial ladder. For the first time, small towns and cities could now afford the luxury of an aerial ladder truck. *Pirsch Junior* aerial ladders were even mounted on some competitive makes – including American-LaFrance.

In 1934, American-LaFrance strengthened its standard spring-raised wooden aerial ladder, add-

Successor to venerable American-LaFrance *Type 31* was pug-nosed *Master Series Type 231 Front-Drive Aerial* introduced in 1930. Only five were built between 1930 and 1932. Rapid City, South Dakota, got this 75-foot *Type 231* in early 1931. *WMC*

ing heavier stanchions and steel-pipe handrails to the base section, permitting the ladder truck to do double-duty as a water tower. A three-inch hose line was strapped to the underside of the bed section and connected to a permanently mounted ladder pipe at its tip. Elevation and nozzle stream direction were controlled from the ground by a series of guy ropes and pulleys.

Major advances in aerial ladder design came thick and fast during the second half of the decade. In the spring of 1935, the Peter Pirsch & Sons Co. delivered the nation's first *all-metal, 100-foot, hydraulically operated* aerial ladder truck to Melrose, Massachusetts.

"Pirsch again leads the field with the introduction of this 100-foot all-metal ladder," the company proclaimed in its announcement ad in fire service trade publications. *"For the first time, an American-built aerial ladder longer than 85 feet is now available. It is all-powered and 100-per cent controlled by one man. We offer it after two years of development work and study,"* Pirsch boasted.

Drawn by a Pirsch custom-chassis tractor, the bridge-like aerial ladder had high handrails on all three sections. The tillerman sat low in a flip-over type seat nested in the ladder, rather than sitting on top of it. This milestone aerial ladder truck remained in service in Melrose into the 1970s, and is believed to be still in existence today.

The Seagrave Corporation also introduced some important product advances in 1935. Not content to let rival American-LaFrance hog the spotlight with its industry-leading V-12 engine, Seagrave announced its *own* 900 cubic inch V-12 fire service engine in 1932. Because of the depression, however, the 240-horsepower engine did not go into production for another three years, with deliveries commencing in early 1935. Seagrave's big "J" V-12 was showcased in a handsome new front-end ensemble with tall "sweetheart" grille.

But Seagrave's *big* news that year was the industry's first all-steel service aerial ladder truck. Under development for several years, the Seagrave service aerial was equipped with a 65-foot hydraulically operated, three-section all-steel aerial ladder. The first of these stylish, low-slung service aerials was delivered to Lancaster, Ohio. Today – in remarkably original

German fire apparatus manufacturer *Magirus* introduced European-style *rearmount* aerial to American fire service in 1924. But compact, rearmount configuration was slow to catch on. *American Magirus Co.* sold fewer than a dozen to U.S. and Canadian fire departments (seven in U.S., four in Canada) between 1925 and 1932. This is one of two Magirus 100-foot aerials delivered to Montreal, Quebec, in 1931. *WMC*

condition – this historic piece of fire apparatus is on display in the Columbus Fire Museum in Columbus, Ohio. Seagrave delivered its first 75-foot tractor-drawn hydraulically operated steel aerial to Pasadena, California, in 1936. The company made its first 85- and 100-foot three-section steel aerials the following year. In 1937 and 1938, Seagrave delivered three 100-foot aerials with closed four-door sedan type cabs to Detroit. Los Angeles and Cleveland, Ohio, later also received aerials of this advanced design.

In early 1937, Peter Pirsch & Sons began production of the distinctive, close-latticed, *aluminum-alloy* aerial ladder that would become this company's most recognizable product for the next 50 years. The tractor-drawn, three-section 85- or 100-foot ladder was of *riveted* – rather than welded – construction. In its advertising, Pirsch claimed that the new 100 percent hydraulically operated metal aerial ladder weighed only a *third* as much as a welded steel ladder and just *two-thirds* as much as a wooden one, but was *stronger* than either.

"Time-tested methods of fabrication are used in the construction of the Pirsch Metal Ladder. Rivets have been used for years in tall structures, derricks, bridges, trestles etc. The Pirsch ladder is the strongest and most durable built. It will not rust and thereby weaken the ladder. The submerged tiller seat (with windshield) is the only improvement ever made with the thought in mind of guarding the tillerman's life," the company proclaimed.

Within a few years, Pirsch had expanded its aerial truck product line with 65- and 75-foot service versions of its lattice-type metal alloy ladder. The company now offered a whole family of aerials consisting of the 55-foot *Pirsch Junior;* 65- and 75-foot *Pirsch Intermediate* service aerials, and tractor-drawn 85- and 100-foot *Pirsch Senior* aerials.

New York City – the nation's largest metropolitan fire department –continued to order new aerial trucks in impressive numbers: ten American-LaFrance *Type 217 Master Series* 75-foot tractor-drawn aerials in 1931; five Combination Ladder Co. 75-foot aerials with pug-nosed Walter tractors in 1933; eight more Walters with Pirsch 75-foot aerials the following year; and ten Seagrave 85-foot tractor-trailer aerials in 1937. In the mid-1930s, the *Four Wheel Drive Auto Company* (FWD) of Clintonville, Wisconsin – which had built many four-wheel tractors for the FDNY in the 1920s – started building complete tractor-drawn aerials with two-section, spring-raised wood ladders. The FDNY purchased no fewer than 26 of these – twelve in 1936 and fourteen more in 1938.

Industry leader American-LaFrance, meanwhile, wasn't exactly sleeping at the switch while all this was going on. The Elmira, New York, firm had been quietly developing its own "next-generation" aerial ladder products. With the Depression apparently finally receding, the company announced three major advances in 1938:

- The company's first *hydraulically operated* aerial ladder.
- The company's first *all-steel* aerial ladder
- The industry's first compact, *cab-forward* service aerial

In early 1938, American-LaFrance delivered its first hydraulically operated aerial ladder truck to Union City, New Jersey. Pulled by a sleek 400 Series tractor, the Model 400-TB had a *three-section* 85-foot wooden aerial ladder – the only one of this configuration ALF ever made. Six months later, the company delivered its first metal aerial to Annapolis, Maryland. Also pulled by a Senior 400 Series tractor, this impressive aerial was equipped a *four-section* 100-foot steel aerial ladder – the company's first.

At about the same time in mid-1938, American-LaFrance introduced its art deco *500 Series* pumpers and aerials. The radically styled, all-new 500 Series fire apparatus was as modern as tomorrow. The big 500 sported a massive sloped grille with prominent center split and a series of fine horizontal grille bars. The 500's ultra-streamlined front-end ensemble featured a long, wide hood, deep cowl and narrow slit-like windshield with small triangular quarter windows. On pumpers, suction hoses, ladders and all other equipment were stored inside the 500's sleek, all-steel "envelope" body. Even the bell was tucked out of sight behind the 500's spacious grille cage. A choice of two V-12 engines was available.

The new 500's bold styling was extended to all other models. The new line included pumpers, city service ladder trucks and quadruple combinations, straight-frame service aerials and quints – and tractor-drawn aerial ladder trucks. Newark, New Jersey, got the first 500 Series tractor-drawn *Rescue Series* aerial – a huge white-painted 100-footer with open semi-cab. These aerials were available with a three-section 85-foot or four-section 100-foot all-steel, hydraulically operated aerial ladder.

A notable design change on ALF's new metal aerials was a *fixed* tiller seat, windshield and steering wheel behind the nested ladder. It was no longer necessary to remove the windshield, steering wheel and tiller seat to raise the aerial ladder, which was controlled from a modern pedestal on the broad turntable. Two-section, spring-raised 75- and 85-foot wood aerial ladders were also still available in the new series.

But American-LaFrance's *tour de force* was the truly revolutionary cab-forward *JO/JOX* service aerial, also introduced in 1938. The straight-frame *Model JO and JOX* were the first *compact, cab-forward* fire trucks built in America. The trend-setting JO/JOX established the basic design configuration for service aerial ladder trucks for many years to come. Alfco's breakthrough design placed the engine – and driver – out in front *ahead* of the front axle. An extremely wide sheet-metal cab enveloped the entire front end of the vehicle. Designed by engineer John Grybos who had joined ALF in 1927, the JO/JOX featured a massive, full-width cab with two-section mesh grille. Separate bucket seats for the driver and officer flanked each side of the wide engine compartment. A 190-horsepower V-12 engine was standard.

Unlike long tractor-trailer aerials then in use in most U.S. cities, compact *Magirus* rearmount took up about same space on street as a pumper. Here's one of Montreal's Magirus 100-foot aerials in action as a water tower at a downtown fire in the 1940s. Note tarp over open cab and Bickle city service ladder truck in foreground. *WMC*

The groundbreaking JO/JOX was nearly a third shorter than a conventional tractor-drawn aerial ladder truck of the day. The new aerial's overall length of just under 42 feet meant that it would fit into almost any fire station with an apparatus bay intended to house a city service hook and ladder truck. And there was no need for a tillerman to steer the rear end.

The stunningly modern-looking four-wheel type service aerial was built in two basic styles. The standard *JOX* sported conventional arched rear fenders and an open double-bank ladder bed. The very stylish, top-line *JO* had fully enclosed ladder bed and streamlined "teardrop" rear fenders. The JOX was equipped with a pair of manually operated stabilizer jacks which swung out from the sides of the vehicle below the turntable. The deluxe JO had hydraulically operated stabilizers that extended straight down to the ground. All JO/JOX aerials were of the open, semi-cab style. Hydraulically operated three-section steel aerial ladders of 65-, 75- and 85-foot lengths were initially offered. A four-section 100-foot ladder became available two years later.

The first of these ultramodern-looking cab-forward service aerials was delivered to Marietta, Ohio, in the summer of 1938.

Until the arrival of the JO/JOX, American-LaFrance had not built a straight-frame aerial in about six years. Just half a dozen of its front-drive Master Series Type 231 aerials had been sold between 1930 and 1932. In 1939, the company introduced its first *500 Series* service aerials. The new *Model 190* had a midship-mounted 65-foot three-section steel ladder which could easily be operated by one man.

As the 1930s drew to a close, the hydraulically operated, all-steel aerial ladder was rapidly gaining favor with America's fire chiefs. With its high side rails from turntable to tip, the steel ladder offered an added degree of safety for fire fighters – and the public – in rescues from upper floors. The rigid steel ladder could also be used as a water tower, with a portable ladder pipe attached to its uppermost rungs, or under the bed ladder. Almost immediately, the versatile *Combination Aerial Ladder/Water Tower* marked the end of factory-built water towers in America – the last of which was built in 1938.

American-LaFrance, Seagrave, Pirsch, FWD and Ahrens-Fox competed vigorously for aerial ladder sales. Mack Trucks, however, had ceased making aerial trucks after 1935. The *Maxim Motor Company* built its first two service aerial ladder trucks – with 65-foot, three-section steel ladders and closed cabs – in 1939. In addition to their new steel aerials, American-LaFrance and Seagrave continued to offer their customers spring-raised wooden aerials. Ahrens-Fox was building fewer and fewer of its pneumatically raised wood aerials, but in 1939 came out with a new steel-reinforced *Bascule* aerial – one of which was mounted on a gigantic piston-pumper quintuple combination for Greenburgh, New York.

Peter Pirsch & Sons introduced America's first *fully powered* aerial ladder truck in 1931. The two-section 85-foot wood aerial was raised, lowered and rotated entirely by Pirsch-designed hydro-mechanical hoist. The prototype, drawn by a White heavy-duty tractor, was delivered to Spokane, Washington, on June 5, 1931. *MLC*

Pirsch — not American-LaFrance or Seagrave — pioneered the *service aerial* — a city service ladder truck equipped with an aerial ladder. Peter Pirsch & Sons of Kenosha, Wisconsin, introduced the *"Pirsch Junior Aerial,"* a light, manually operated 55-foot aerial ladder — in 1931. Designed for mounting on any city service ladder truck, the midship-mounted Pirsch aerial was easily operated by one man. Pirsch was soon delivering "junior aerials" on its own custom chassis. Among the first was this one — with pump — for North Arlington, New Jersey. *MLC*

1932 Pirsch *Model 15 Junior* Aerial Ladder Truck built for Billings, Montana. *MLC*

Almost overnight, Peter Pirsch & Sons in the early 1930s became a leading manufacturer of aerial ladder trucks. Minneapolis, Minnesota, took delivery of this Pirsch 85-foot *Senior* tractor-drawn aerial with hydro-mechanical hoist in early 1932. *MLC*

In a convincing demonstration of its strength, eleven firefighters pose for the photographer on the prototype Seagrave steel aerial ladder in downtown Columbus, Ohio, in 1935. One man could easily raise, rotate and lower the 65-foot steel aerial ladder from a control pedestal on the left side of the truck's turntable. *MLC*

Not to be outdone by rival Pirsch, Seagrave introduced the industry's first 100 percent hydraulically operated, all-steel *Service Aerial* in 1935. Built on a standard Seagrave city service ladder truck chassis, its breakthrough design featured a 65-foot, three-section steel aerial ladder. Still on the road today, the first one was delivered to Lancaster, Ohio, in 1935. *WMC*

In 1934, American-LaFrance substantially strengthened its spring-raised wood aerial ladder with the addition of heavier tubular steel handrails and stanchions on the base, or bed, section. Lynn, Massachusetts, got this V-12 powered *Type 317* tractor-drawn 85-foot aerial in 1935. The last of these quick-raising *American Automatic* aerials was built in 1950. *JFS/WMC*

At the depths of the Great Depression, some cities actually built their own fire apparatus — including aerial ladder trucks. Jersey City, New Jersey's resourceful shops constructed no fewer than eight aerial ladder trucks, including this one with 1935 GMC tractor and American-LaFrance 75-foot wood aerial. *JFS/WMC*

Seagrave introduced its first V-12 engine in 1935 — four years after American-LaFrance. The 900-cubic-inch engine was showcased in a handsome new front-end ensemble with high, tapered hood and stylish "sweetheart" grille. Boston, Massachusetts, placed this Seagrave V-12 tractor-drawn 85-foot aerial in service in 1936. *Bill Noonan Collection*

The *Four-Wheel-Drive Auto Co.* (FWD) of Clintonville, Wisconsin, began making complete tractor-drawn aerial ladder trucks in the mid-1930s. New York City purchased 12 of these FWD spring-raised 85-foot wood aerials in 1936, and 14 more two years later. *Robrecht Negatives*

Peter Pirsch & Sons racked up yet another first in American aerial ladder evolution with delivery of the industry's first *all-powered, 100-foot metal aerial* in 1935. Built for Melrose, Massachusetts, the milestone $17,000 aerial had a three-section metal ladder with high handrails on each section. Note also the bed of metal ground ladders. *Murray Young*

Pump-equipped 1937 *Pirsch Junior* aerial originally built for Niles Center, Illinois (later, Skokie) and later sold to Westchester, Illinois. This one has a manually operated *Pirsch Junior* 55-foot ladder. Some were delivered with electric hoists. Note stacked siren, bell and Mars light in front of new V-type radiator — a Pirsch hallmark. *Dan G. Martin*

Pirsch built the first of its famed close-lattice aluminum-alloy aerial ladders in 1937. *Pirsch Senior* aerials were available in three-section 85- and 100-foot sizes. Peter Pirsch & Sons produced this very distinctive ladder design for half a century, into the 1980s. Three of the company's first 100-foot aerials of this design went to Washington, D.C., in 1937. *MLC*

Despite ravages of the Great Depression, vehicle styling made tremendous advances during the 1930s. Compare this streamlined, sedan-cab tractor-drawn 100-foot aerial with one built only a few years earlier. This state-of-the-art aerial was delivered to Detroit Fire Department in 1938. Note chemical equipment on trailer and triple stack of pompier ladders. *WMC*

Introduced in 1935, Seagrave's versatile 65-foot *Service Aerial* — essentially a city service ladder truck with an aerial ladder — quickly became one of the company's most popular products. The hydraulically operated, three-section 65-foot steel aerial was nestled between two full-length hose boxes. This one was built for Phoenixville, Pennsylvania, in 1938. *MLC*

Industry giant American-LaFrance began experimenting with hydraulic hoists and metal aerial ladders in the mid-1930s. The company's first hydraulic hoist was installed on this *Senior 400 Series* tractor-drawn aerial delivered to Union City, New Jersey, in mid-1938 — the *only* aerial ALF ever built with a *three-section* wood ladder! *WMC*

Following introduction of its 65-foot service aerial in 1935, Seagrave successively introduced 75-, 85- and 100-foot hydraulically operated all-steel aerial ladders in the 1930s. The first two 100 footers (and the very last one in 1968) went to Detroit, Michigan. Here's one of the two 100-foot tractor-drawn aerials with fully enclosed four-door safety sedan cabs delivered in 1937, this one assigned to Ladder 1. *MLC*

American-LaFrance launched its dramatically styled art deco *500 Series* in June 1938. At the top of the 500 Series product line was the new *Rescue Model* tractor-drawn aerial with hydraulically operated 100-foot steel aerial ladder. The first 500 Series aerial delivered was this massive white-painted 100 footer for Newark, New Jersey. Note vertical lift cylinders on turntable, and fixed tiller seat. *WMC*

High-styled new American-LaFrance *500 Series* pumpers and ladder trucks were available with a wide choice of open or closed cabs. This *500 Series* 100-foot aerial with closed cab was shipped to Dearborn, Michigan, in late 1939. Note whirling *Buckeye Roto-Ray* warning light on cab roof and bell concealed behind art deco grille. The author photographed this big aerial late in its fire service career in 1963. *WMC*

An impressive public demonstration of the first American-LaFrance *500 Series* aerial was held in downtown Elmira, New York, before the huge rig was shipped to Newark, New Jersey, in mid-1938. The hydraulically operated four-section 100-foot steel aerial ladder was stretched to the roof of the eight-story Mark Twain Hotel. *WMC*

Important design feature of new American-LaFrance *500 Series* tractor-drawn aerials was fixed tiller seat. For the first time, the tillerman's windshield, seat and steering wheel did not have to be removed when the aerial ladder was to be raised, saving precious seconds — vital in rescue work. This 1939 "500" has three-section 85-foot aerial. *MLC*

Tidy turntable on 1939 American-LaFrance *500 Series Rescue Model* tractor-trailer aerial ladder truck with three-section 85-foot ladder. Note control pedestal on right and angled hydraulic lift cylinders on each side of bed ladder. *WMC*

The face of the future! American-LaFrance revolutionized aerial ladder truck design in America with introduction of its compact, cab-forward *JO* and *JOX* service aerials in 1938. Driver and officer were positioned *ahead* of front axle. Standard 190-horsepower ALF V-12 engine was located between them. Here's a close up of the art deco front end of one of the first of these pioneering aerials delivered to Oneida, New York, in June 1938. *MLC*

Compact, cab-forward American-LaFrance JO/JOX service aerials were miles ahead of anything else on the market at the time. Advanced yet functional styling was as practical as it was daring — frequently compared to futuristic streamlined passenger trains of the day. Striking all white 1938 *Model JO* 85-foot aerial with enclosed ladder beds and teardrop rear fenders was built for Union City, New Jersey. *WMC*

Industry-leading American-LaFrance cab-forward aerials were offered in two versions — streamlined, premium *JO* series with enclosed ladder bed and teardrop rear fenders, and plainer, more utilitarian *JOX* shown here with open ladder racks and conventional rear fenders. This is a 1939 JOX with 85-foot, three-section aerial ladder. *MLC*

Complementing cab-forward *JO/JOX* service aerials introduced the previous year, American-LaFrance introduced a more conventional *500 Series* junior service aerial in 1939. New *Model M-190* was initially available with only 65-foot, three-section aerial ladder. Note vertically acting hydraulic jacks under turntable. First Type 190 went to Rumford, Maine. *WMC*

Maxim Motor Company of Middleborough, Massachusetts, built its first two aerial ladder trucks in 1939 — both for the U.S. government. Maxim didn't begin civilian aerial truck production until after the Second World War. With closed cab, pump and three-section 65-foot steel aerial, this one went to Advance Base Depot in Davisville, Rhode Island. Twin went to U.S. Naval Air Station, Miramar, California. *Fred W. Crismon*

Ahrens-Fox introduced sturdy *Bascule Aerial* in the late 1930s. One of these three-section 70-foot aerials with hydraulic hoist was installed on this monstrous quint with 1000 gpm front-mounted piston pump built for Fairview Fire Company in Greenburgh, New York, in 1939. Ahrens-Fox delivered its last aerials in 1940. *WMC*

Memco Aerial Ladder Co. of Oklahoma City, Oklahoma, was a major manufacturer of light-duty aerial ladders for public utilities services. Inevitably, some *Memco* aerial ladders wound up on fire apparatus — like this home-built ladder truck with Darley front-mount pump and Memco 65-foot metal aerial on 1938 Ford cab-over-engine chassis. Well-known fire apparatus photographer Shaun Ryan found this unusual rig in Hot Springs, South Dakota. *Shaun P. Ryan.*

Despite increasing popularity of all-powered steel aerials, some fire departments still preferred spring-raised, wood aerials. This 75-foot *500 Series* tractor-drawn aerial with open semi-cab was delivered to Schenectady, New York, in September 1940. Note tiller windshield and ladder pipe permanently attached to base section of main ladder. *WMC*

CHAPTER FIVE: THE 1940s

Just as the Great Depression overshadowed all else during the "dirty thirties," America's involvement in the Second World War profoundly impacted events in the first half of the 1940s. The 1940s witnessed the swift adoption by the North American fire service of the hydraulically operated, steel aerial ladder over the spring-raised wooden aerial ladders used since the turn of the century.

As the 1940s began, a distinctly new *type* of aerial truck arrived on the scene. Popularized at first by Seagrave, the *quintuple combination* or "quint" was essentially a quadruple combination city service ladder truck, or *quad*, equipped with an aerial ladder – which further enhanced the rig's already-impressive versatility. A veritable fire department on wheels, the *quint* incorporated five basic firefighting functions on a single chassis – a full-sized 500-, 600- or 750-gallon-per-minute fire pump; booster tank, pump and hose; a full complement of portable ground ladders ranging from 10 to 55 feet; an aerial ladder; and a full-length hose bed for 2-1/2" hose. Most prewar quints had 65-foot, three-section steel aerial ladders. Longer aerial ladders and higher capacity pumps were gradually introduced after the Second World War.

In 1940, American-LaFrance added a 100-foot steel aerial ladder to its cab-forward JO/JOX service aerial family which, since its introduction two years earlier, had only been available with 65-, 75- or 85-foot three-section ladders. Despite the added weight of the four-section, 100-foot ladder, the JO/JOX was

Until 1940, cab-forward American-LaFrance *JO/JOX* aerials were available only with three-section 65-, 75- or 85-foot ladders. By 1940, a four-section 100-foot aerial had been added to the line. Only one engine, a 190-horsepower ALF V-12, was available. All JO/JOX aerials had open semi-cabs. This 1940 *Model M-5-100-JO* was built for Terre Haute, Indiana. *MLC*

still available with just one engine – ALF's Lycoming based "small" 190 horsepower V-12.

Also in 1940, Ahrens-Fox delivered its last aerial ladder trucks – four tractor-drawn 85-footers with special spring-raised hoists instead of the company's signature Dahill air hoist – to the New York City Fire Department (FDNY). The famed Cincinnati manufacturer had produced more than 70 six-wheel, tractor-drawn aerials over a 17-year period between 1923 and 1940.

With war clouds gathering, industry leader American-LaFrance announced a new series of heavy-duty "super aerials" in 1941. These humongous tractor-drawn 500 Series aerials included the *Model HD-125* – with a five-section, 125-foot steel aerial ladder – and the even larger *HD-150* which was supposed to have a six-section, 150-foot aerial – the longest ever made in America. The 125-foot version was equipped with the 500 Series *Rescue Model* aerial's fixed tiller seat and windshield. With an overall length of more than 54 feet, the monstrous HD-150 had a flip-over style tiller seat and windshield nested in the top section of the bridge-like ladder.

Just one of these skyscraper aerials was actually built and placed into service. Delivered to Boston, Massachusetts, in August of 1941, the huge, white-painted Model HD-125 with three-man closed cab tractor had a short, tragically troubled fire service career. Just a month after Boston's first steel aerial was assigned to the B.F.D.'s Ladder Company #8, the truck's five-section steel ladder mysteriously buckled during an evening drill in Fort Hill Square. The grotesquely twisted ladder had to be cut up with acetylene torches the following day before the big truck could be moved. Repaired by American-LaFrance and returned to service, the huge white rig was buried under a falling wall at a major fire the following year at which six Boston firefighters were killed. Again rebuilt, repainted red and retrofitted with a more conventional 100-foot, four-section ladder, and later reassigned to the B.F.D.'s Ladder 19, the rig's apparently jinxed life came to an abrupt end in 1947 when the truck overturned on a test run, killing a member of its crew. Long nicknamed "the *White El-*

Peter Pirsch & Sons introduced contemporary new front-end styling in the late 1930s. With subtle changes, this same basic styling continued into the 1960s. This 75-foot aerial with open semi-cab and enclosed ladder beds was built for Vancouver, Washington, in 1940. Note two-piece "V"-type windshield and shiny bell behind grille. *MLC*

Ahrens-Fox Fire Engine Co. delivered its last aerial ladder trucks — four special *Model W-85s* with spring-raised hoists — to New York City in 1940. This fabled company had produced more than 70 tractor-drawn aerials — and just *one* straight-frame service type — since 1923. *MLC*

ephant," the ill-fated 1941 American-LaFrance was immediately retired and scrapped.

Over the ensuing 18 months, two *more* American-LaFrance steel aerials – both four-section, 100-footers – also structurally failed, further tarnishing the Elmira, New York, manufacturer's reputation. Two months after the initial Boston incident, Dayton, Ohio's brand-new 100-foot ladder unexplainably buckled while being raised to the roof of a downtown department store. A huge crowd witnessed this embarrassing spectacle, as Santa Claus himself was supposed to climb the showpiece ladder to officially begin the city's Christmas shopping season.

More tragically, a Salt Lake City, Utah, firefighter was killed and three others injured when that city's relatively new ALF 100-foot steel aerial ladder twisted and crumpled to the ground at a hotel fire in the city's downtown area in June 1943. The four firefighters were climbing the ladder to the hotel's ninth floor when it suddenly crumpled, throwing them to the ground. All three incidents were later determined to have been caused by metallurgical flaws and suspect welding.

Such embarrassing incidents were not limited to industry leader American-LaFrance. The top section of a 65-foot Seagrave service aerial had failed in Evanston, Illinois, in 1940. These incidents undoubtedly created a degree of mistrust among conservative fire chiefs – more than a few of whom chose to stick with the old style spring-raised, wooden aerials long after the introduction of hydraulically operated metal ladders.

The Chicago Fire Department placed its first steel aerials – two tractor-drawn 100-foot American-LaFrance 600 Series rigs with three-man closed cabs – into service in 1942.

Immediately following the Japanese attack on Pearl Harbor in December 1941, the United States was thrust into the war that had been raging in Europe since 1939. Overnight, U.S. industry shifted to all-out war production. The last civilian vehicles rolled off Detroit assembly lines in early 1942. Americans had to learn to "do without." Suddenly, municipalities still recovering from the effects of the Depression could no longer order new fire apparatus. Allocation and production of new fire equipment was strictly rationed by the War Production Board. Government

Seagrave introduced this pleasing new front-end styling in 1936. Here's a conventional 65-foot service aerial with open semi-cab built for Concord, Massachusetts. Note the one-piece, fold-down windshield and Seagrave *Twin Blast* siren mounted on the cowl. *MLC*

Seagrave Corporation of Columbus, Ohio, built its last spring-raised wood aerial ladder trucks in 1941. The final order was for seven 75-foot tractor-drawn aerials with doorless open cabs for New York City. Note tiller windshield. Ready to ship, this one was photographed on South High St. across from the Seagrave plant. *MLC*

and military requirements got top priority, of course, followed by municipalities in which vital defense industries or military installations were located. Raw materials were also severely rationed, resulting in very drab-looking fire apparatus with painted grilles, bumpers, hardware and no brightwork or fancy gold leaf trim of any kind.

America's fire apparatus manufacturers made a significant contribution to America's war effort. Between 1942 and 1945, American-LaFrance, Seagrave, Pirsch, Maxim, Ward LaFrance, Mack, FWD and many smaller manufacturers like Howe, Boyer and Darley turned out thousands of fire trucks, aircraft and gun parts and other war materiel. These wartime products ranged from civil defense-type trailer pumps, warehouse hose reels and wheeled chemical fire extinguishers to military-spec fire trucks, airfield crash tenders and special motor fire apparatus to protect shipbuilding, petroleum and chemical industries.

During the war, relatively few aerial ladder trucks were built by American-LaFrance, Seagrave and Pirsch for the U.S. Government, primarily for the protection of major military installations like supply depots and shipyards. In 1942, the U.S. Treasury Department ordered ten American-LaFrance JOX 100-foot service aerials which were exported to far-off Australia under the government's Lend-Lease Act. Several of these big aerials are still on the road "down under" today.

Perhaps the most interesting proposal put forth early in the war was an ALF proposal for a much-modified JOX aerial with a battery of anti-aircraft guns at its tip! This bizarre proposal never made it past the blueprint stage.

By 1945, with victory assured, military contracts dried up and fire apparatus manufacturers began to prepare for the resumption of peacetime production.

The postwar period triggered one of the biggest sales booms the industry had ever experienced. Fire departments across the country, which had been unable to buy new fire apparatus during the Depression, scrambled to place orders for new pumpers, ladder trucks, and hose and rescue units to replace the now-obsolete, worn-out rigs with which they had had to make do since the 1920s. By 1946, the assembly lines were humming again and everything the factories could produce was eagerly snapped up by a long waiting list of customers. Even then, production was frequently hampered by strikes and material shortages.

In the main, fire apparatus manufacturers answered this huge pent-up demand with what were essentially warmed-over prewar fire apparatus designs. Development of all-new postwar fire apparatus designs would simply have to wait.

But there was *one* spectacular exception. At the 1943 annual convention of the International Association of Fire Chiefs (IAFC), the American-LaFrance Foamite Corporation conducted an in-depth survey to determine what features America's fire chiefs

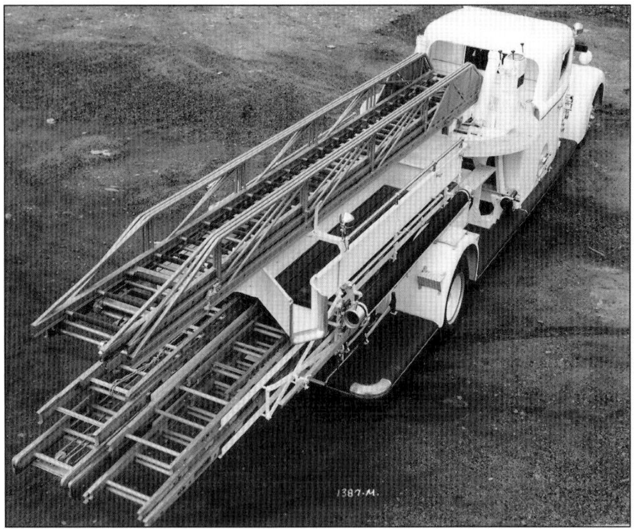

High angle rear view of a 1941 American-LaFrance *M-190* Quadruple Combination with closed-canopy-style cab delivered to Streator, Illinois, in 1941. Note "Y"-shaped hose playout trough under three-section 65-foot aerial ladder, hard suction hose and full complement of wood ground ladders. *MLC*

American-LaFrance introduced plainer, narrow-cab *Type 500* in 1940. The Philadelphia Bureau of Fire purchased nine of these *Type 500* tractor-drawn aerials with 85-foot spring-raised wood ladders in 1940 and 1941. Note the post-mounted *Roto-Ray* warning light on left side of the wide-open doorless cab. *Robrecht Negatives*

The next significant trend in aerial truck design was the appearance of the *Quintuple Combination* in the early 1940s. The versatile *Quint* combined five basic firefighting functions on a single chassis — pump, booster water tank system, hose bed, a full complement of ground ladders, and an aerial ladder. This well-equipped Seagrave Quint was delivered to Bryn Mawr, Pennsylvania, in 1941. *MLC*

Seagrave pioneered the rear-entrance, *canopy-style cab* in 1937. Four firefighters sat on a rear-facing bench seat under an extended rear roof canopy. The new safety cab design got men off the running boards and under cover. New Haven, Connecticut, ordered this 75-foot tractor-drawn aerial with canopy cab tractor in 1941. The one-piece bucket-type tiller seat flipped out of the way when the main ladder was to be raised. *MLC*

would like to see in postwar fire apparatus design. The 28-page *Design Symposium* booklet contained 24 questions concerning basic design and configuration. Respondents were asked about driver visibility, safety features and equipment placement and storage.

But the real purpose of this slick little booklet was to prepare the American fire service for the pending arrival of the company's revolutionary *cab-ahead-of-engine* fire apparatus, which was already in the final design phase in Elmira, New York. In 1944, several scale models of the proposed new cab-forward design were completed for evaluation and study. The styling models included a pumper with open semi-cab, a four-wheel service aerial and – remarkably – a short-wheelbase aerial truck with a rear-mounted ladder.

Successors to the company's prewar 500 and 600 series, these bold, new products were christened the *700 Series.* Two full-sized "pilot" models were completed in 1945 – a 1500-gallon-per-minute canopy cab pumper and a 100-foot service aerial with open semi-cab. The new 700 Series was announced with much fanfare in the October 1945 issue of *Fire En-*

gineering, at the time the U.S. fire service's leading trade journal.

American-LaFrance's truly revolutionary new 700 Series pumpers and aerial ladder trucks placed the driver ahead of the engine, with vastly improved forward visibility compared to the company's prewar designs. The V-12 engine and radiator were located behind the front seat and axle. A pair of rear-facing jump seats flanked the engine compartment. While the cab-forward pumper represented a total new concept in pumper design, the new 700 Series service aerials were essentially an improved, updated evolution of the pioneering JO/JOX four-wheel aerials introduced in 1938. In the wide-cab JO/JOX, however, the engine was placed between the two front seats, behind a conventional grille.

The new 700 had a tapered, smooth-faced cab front with no grille. With its chrome center windsplit and angled two-piece windshield, this very modern-looking front end was somewhat reminiscent of General Motor's futuristic "F" series locomotive – the *Train of Tomorrow.* Customers could specify a five-man semi-open cab or closed canopy-style cab. Unlike the prewar JO/JOX, the new 700 Series aerials were available with a choice of two V-12 engines – a 215-horsepower "J" V-12 or the big 275-horsepower Model "E" motor.

The 700 Series aerial truck lineup offered four-wheel service aerials with three-section 65-, 75- or 85-foot metal aerial ladders, or a four-section 100-foot ladder. At the top of the comprehensive 700 line were the six-wheel, tractor-drawn *Rescue Series* aerials with 85- or 100-foot ladders. Quintuple combination aerials with variously sized pumps and aerial ladder lengths were also available.

The last 600 Series fire engines and JO/JOX aer-

American-LaFrance built this huge *500 Series* tractor-drawn aerial with four-door sedan type cab for Cleveland, Ohio, in 1941. Note the Federal C5 siren on hood, handrail behind the cab and new box-style fenders on trailer. Tillerman's windshield, seat and steering column did not have to be moved to raise four-section, 100-foot steel aerial ladder. *WMC*

Industry leader American-LaFrance announced two new *Super Aerials* in 1941. Tallest in the industry at the time, they included five-section, 125-foot *Model HD-125* with fixed tiller seat, and six-section, 150-foot *HD-150* with flip-over type tillerman's seat. Only one HD-125 was ever sold. As far as is known, *no* 150 footers were ever built. *WMC*

Boston Fire Department got the only 125-foot American-LaFrance aerial ever built. Huge, white-painted rig with five-section ladder had a short, tragic career. Showpiece ladder collapsed during drill shortly after it was placed in service. Repaired rig was heavily damaged by a falling wall in 1942. Three years later a crewmember died in a rollover accident. Rig was then scrapped. *WMC*

American-LaFrance and Seagrave continued motor fire apparatus production through the Second World War, with bulk of production going to U.S. government and military. Seagrave shipped this 85-foot service aerial, a pumper and a booster hose truck to Pearl Harbor in 1943. Note olive-drab paint job and total absence of any brightwork. *MLC*

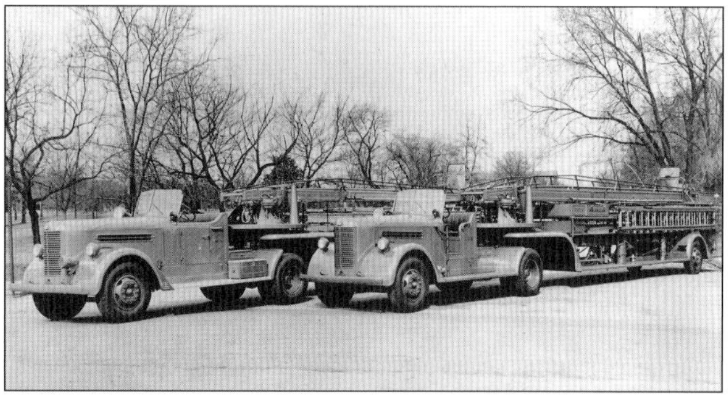

A pair of Pirsches. Nashville, Tennessee, received this matched set of Peter Pirsch & Sons 85-foot, spring-raised wooden aerials in 1943. Despite popularity of the company's aluminum alloy aerial with hydro-mechanical hoist, some fire departments still preferred the older style wood aerials. *MLC*

ials were produced in the latter part of 1946. The sprawling Elmira, New York, plant was then completely retooled to produce the daring new 700 Series exclusively. The first 700 Series pumpers and aerials came off the Elmira assembly line in the first quarter of 1947. The first 700 Series aerial delivered to a fire department was a canopy-cab 85-foot service aerial delivered to Binghamton, New York, in early April 1947.

The company's understandably nervous competitors were quick to ridicule the cab-forward design, calling it the "suicide" cab and yet another example of form over function. But the breakthrough 700 Series was accorded an overwhelmingly favorable reception by the fire service. All other U.S. fire apparatus builders would eventually adopt the cab-ahead-of-engine design. Today, well over half a century later, nearly all of the custom-chassis fire apparatus built in North America is of precisely the same cab-forward configuration American-LaFrance pioneered with its daring departure in fire apparatus design.

In a historical footnote, the very first American-LaFrance 700 Series fire apparatus completed is still on the road today. Constructed in 1944-45 and assigned factory serial number *9000,* the 100-foot service aerial with open semi-cab was used as a factory demonstrator and rolling test bed for eleven years before it was sold to Jamestown, New York, in 1956. Rebuilt in 1980, the very first member of this long, proud line later saw service with fire departments in Falconer and Cattaraugus, New York, before it was finally retired in the 1990s.

Although the Maxim Motor Company of Middle-

Pirsch aerials were made in three distinct series: 50- and 55-foot *Pirsch Junior,* 65- and 75-foot *Intermediate,* and 85- and 100-foot *Senior.* Here's a rear view of a 65-foot *Pirsch Intermediate* service aerial with rear entrance canopy-style cab delivered to Milwaukee, Wisconsin, in 1944. Longest portable ground ladder carried is Pirsch trussed 50 footer with "tormentor" poles. *MLC*

borough, Massachusetts, had built its first two aerial ladder trucks in 1939, the company did not add aerial ladder trucks to its product line until 1947. Mounted on the company's handsome new custom conventional chassis, Maxim's aerial offerings initially included 65-, 75- and 85-foot four-wheel service aerials with three- and four-section steel ladders. The first of these was delivered to Milton, Massachusetts, that year.

Up to now, the nation's largest and busiest metropolitan fire department – New York City – had purchased nothing but spring-raised wooden aerial ladders since the turn of the century. But in 1948, the FDNY began its own in-service evaluation of hydraulically operated metal aerial ladders. Three 85-foot tractor-drawn aerials were purchased from three of the leading manufacturers – American-LaFrance, Seagrave and Pirsch – and placed into service at three busy ladder companies. All had open semi-cabs. The ALF 700 Series *Rescue Model* tillered aerial was assigned to Hook & Ladder Co. 24; the wide-cab Seagrave to Ladder 110 and the Pirsch with its distinctive lattice-type ladder to Ladder 4. All had three-section ladders and were equipped with ladder pipes which could be attached to the underside of the base section or the tip of the fly section. In some instances *two* ladder pipes were used on one of these ladders at major fires. With the introduction of these rigid metal aerial ladders with their water tower capability, the decline of the traditional single-purpose water tower began. But another five years would pass before the FDNY made its first large-scale purchase of metal aerials.

By 1949, the Four Wheel Drive Auto Company (FWD) of Clintonville, Wisconsin, had resumed production of spring-raised, wooden aerials. The Chicago Fire Department ordered a dozen FWD 85-foot, tractor-drawn aerials with closed cabs, finished in Chicago's distinctive black-over-red. Many more would follow. Even though the Windy City had acquired its first two metal aerials in 1942, Chicago continued to favor spring-raised wood aerials into the mid-1950s.

Mack Trucks – which had discontinued the manufacture of aerial ladder trucks in the mid-1930s – reentered the booming aerial ladder truck market in 1949. Mack purchased 65- and 75-foot metal aerial ladders from rival Maxim for mounting on its own highly-successful "L" Series fire apparatus chassis.

American-LaFrance built 110 cab-forward *JO/JOX* service aerials between 1938 and 1946 — an even 100 in Elmira, New York, plant and ten more in the company's Canadian subsidiary plant in Toronto. All had open semi-cabs and 190-horsepower V-12 engine. This late-production *JO* with 75-foot aerial was delivered to Hudson, Massachusetts, in early 1945. *WMC*

Seagrave and American-LaFrance built a small number of aerial trucks with special leveling devices for cities — like San Francisco — with hilly streets. This 85-foot Seagrave tractor-drawn aerial with leveling device was built for Phoenix, Arizona, in 1943. Note stationary quadrant at base of ladder. *MLC*

Introduced during the late 1930s, sturdy steel aerial ladders proved ideal for water tower operations. Three American-LaFrance 100-foot service aerials in action at a major fire in Omaha, Nebraska, in the 1950s. Note *two* ladder pipes in operation on the JOX aerial on the left — *and* the old Hale water tower barely visible through the smoke! *MLC*

With the war winding down, American-LaFrance engineers began design work on an all-new postwar *700 Series* in 1943. Design proposals included this rear-mounted 100-foot service aerial. American-LaFrance didn't introduce its first rearmount aerial — of precisely this same configuration — for another 25 years! *WMC*

Until the introduction of the new "B" Series in 1953, all of the postwar Mack aerials were of the four-wheel service type, with open or closed cabs. Mack also delivered some quintuple combination aerials on the popular "L" and "LS" chassis with its classic chrome-plated radiator shell and massive front bumper.

As the twentieth century reached its mid-point the big postwar sales boom was winding down – but unbridled prosperity nonetheless showed every sign of rolling on into the 1950s.

With great fanfare, American-LaFrance announced all-new *700 Series* fire engines in late 1945. Prototype pumper and service aerial were completed in 1944-45 and exhibited at 1946 IAFC Convention. Great-granddaddy of all 700s was this 100-foot aerial, Reg. #9000. Companion 1250 gpm pumper was Reg. #9001. Prototype 700 aerial was sold to Jamestown, New York, in 1956. It's still on the road today. *WMC*

Matched pair of American-LaFrance *JO* and *JOX* aerials owned by Larry Zotti, Granite City, Illinois. At left is his 1942 *M-65-JO Quint*, ex-Sturgis, Michigan (the only one ever made), and 1946 *M-5-100-JOX*, ex-Alton, Illinois. *Larry Zotti Photo*

Prototype *American-LaFrance 700 Series* aerial photographed during testing in 1946. The four-section, 100-foot steel ladder has been extended to its maximum length. Note the manually operated stabilizer jacks visible on both sides of this milestone rig. *WMC*

ALF 700 Series production began in early 1947. The first 700 Series aerial delivered to a fire department was this 85-foot service aerial with closed canopy-style cab built for Binghamton, New York. Note transverse life net compartment and the ladder pipe carried on the running board just ahead of the rear fender. *MLC*

One of the first American-LaFrance 700 Series *Rescue Model* tractor-drawn aerials was this 100 footer for Washington, D.C. Note *inclinometer* on the side of the aerial ladder and lack of fancy striping and trim — as per D.C.F.D. specs. *WMC*

Maxim Motor Company officially introduced its first aerial ladder trucks in 1947. Among the first delivered was this 75-foot service aerial on Maxim's handsome new postwar custom chassis for Milton, Massachusetts. Maxim later sold complete aerial ladder assemblies to rivals Mack and Ward LaFrance. *MLC*

Seagrave introduced a new, wider cab in 1945. This *Model J-85* service aerial with four-section, 85-foot aerial ladder and booster equipment was delivered to Lockport, New York, that year. Note fully enclosed ladder bed and *Mars* wig-wag light perched on nose. *MLC*

1948 *Pirsch Intermediate* 75-foot aerial in action at commercial building fire in downtown Sarnia, Ontario, in 1951. Men in foreground are controlling the ladder pipe with ground lines. Pirsch delivered only three aerials to Canadian fire departments, all within a 50-mile radius in southwestern Ontario — Sarnia, Chatham and Windsor. *WMC*

New Orleans got this Seagrave 100-foot tractor-drawn aerial with open semi-cab in 1947. Three-section steel aerial has been extended to full length. Note the flip-over tiller seat bucket. *MLC*

The last word in Seagrave aerial ladder trucks circa 1948. This imposing 100-foot tractor-drawn job with canopy-style cab was delivered to Manchester, New Hampshire. The long handle on the side of the cab locks the tractor's rear axle. Look closely and you'll also see the road sanders just ahead of the rear fenders on the tractor. *MLC*

After a 14-year hiatus, *Mack Trucks* re-entered the aerial ladder truck market in 1949. Built on the company's handsome "L" Series chassis, Mack service aerials were available with 65- or 75-foot aerial ladders sourced from rival Maxim. Buchanan, Michigan, got this Mack L-Series 65-foot service aerial, with concealed pump, in 1951. *MLC*

Classic American-LaFrance *700 Series* Service Aerial delivered to Benton Harbor, Michigan, in January 1949. Note the handrail behind the front seat and four-section 100-foot aerial ladder. This aerial was powered by ALF's 215-horsepower "J" V-12 engine. *WMC*

Spring-raised wood and newer-style steel hydraulic aerials work side by side at major fire in Bangor, Maine. Ladder pipe on the bed section of the two-section wood aerial is being readied for water tower operation, while firefighters stretch ladder pipeline up the Seagrave service aerial. *WMC*

American-LaFrance delivered its last three spring-raised, wood aerial ladders to the Boston Fire Department in 1950. Originally drawn by ALF 700 Series tractors with booster equipment, the 85-foot wood-aerial trailers remained in service, drawn by a variety of replacement tractors, through the '70s. Ladder 13's rig was photographed on the street when it was still relatively young. *Bill Noonan Collection*

CHAPTER SIX: THE FABULOUS 1950s

This eventful, fondly remembered boom-and-bust decade witnessed the passing of the spring-raised wooden aerial ladder – and the momentous birth of the most revolutionary development in aerial firefighting in America since the invention of the self-supporting aerial ladder itself.

As the 1950s began, just six of the nation's largest fire apparatus manufacturers made and sold aerial ladder trucks – all of them on their own custom or semi-custom fire apparatus chassis.

American-LaFrance Foamite Corporation – Elmira, New York

The Seagrave Corporation – Columbus, Ohio

Peter Pirsch & Sons Co. – Kenosha, Wisconsin

Maxim Motor Company – Middleborough, Massachusetts

Mack Manufacturing Corp. – Long Island City, New York

Four-Wheel-Drive Auto Company – Clintonville, Wisconsin

Two independent makers of light duty municipal utility-type aerial ladders – *Memco* of Oklahoma City, Oklahoma, and *Sponco,* also located in the same city – made "Junior" aerial ladders which were sold in relatively small numbers for aftermarket installation, mainly on commercial chassis fire trucks.

Industry leader American-LaFrance delivered its last spring-raised, wood aerials in 1950. Drawn by 700 Series tractors with open semi-cabs, ALF's last three wood aerials went to Boston, Massachusetts. With two-section 85-foot ladders and *American Automatic* spring hoists, they also carried booster equipment. Long after their original 700 Series tractors wore out, these fast-raising wood aerials served well into the 1970s, hauled by a variety of replacement tractors.

In 1951, the Seagrave Corporation marked its 70th anniversary (1881-1951) with the belated introduction of its first all-new postwar models. The company's *70th Anniversary Series* pumpers and ladder trucks

Peter Pirsch & Sons turned out some interesting bodywork around the three-man cab it purchased from GMC. Here's a fine example with canopy cab extension. This 1950 *Pirsch Intermediate Model 41-D Service Aerial* with three-section, 65-foot ladder was delivered to Knoxville, Tennessee. *MLC*

featured bold, new front-end styling with a massive full-width grille. The siren on these stylish new models was neatly incorporated into the upper nose section of the grille. The first 70th Anniversary Series pumpers and aerials were delivered in mid-1951 to the hometown Columbus, Ohio, Fire Department. With various refinements, Seagrave's long-nose 70th Anniversary Series remained in production for nearly two decades, with the last one built in 1970.

Seagrave's new 70th Anniversary models were available with a choice of V-12 fire service engines. The smaller 400 and 500 Series pumpers and service aerials were powered by a 530-cubic-inch V-12, while big 900 Series high capacity pumpers and tractor-drawn aerial ladder trucks were fitted with Seagrave's 906-cubic-inch "J" V-12. Straight-frame service aerials and quints were available with three-section 65- or 75-foot or four-section 85-foot ladders. Tractor-drawn aerials had three-section 75-, 85- or 100-foot ladders with a flip-over style tiller seat. Customers could specify open, semi-open, three-man closed or rear entrance canopy-style cabs.

Although a German company – *Magirus* – had introduced the short-wheelbase rearmount aerial ladder to the North American fire service in the 1920s, the continental prototype European rearmount aerial never really caught on over here. With fewer than a dozen sales in the U.S. and Canada between 1925 and 1932, *American Magirus* fell victim to the Depression. Ahrens-Fox toyed with a junior-sized rearmount aerial in the late 1930s but this project never got beyond the drawing board. Industry giant American-LaFrance seriously considered a rearmount aerial for its postwar 700 Series, but for some reason backed away from it.

In 1952 the *Maxim Motor Company* of Middleborough, Massachusetts, acquired North American marketing and distribution rights to the highly regarded Magirus aerial ladder, then in wide use around the globe – everywhere, in fact, but North America. Maxim mounted one of these 100-foot four-section hydro-mechanical aerial ladder assemblies on one of its own custom chassis. The result was an extremely compact aerial truck only slightly longer than a pumper. The first of these *Maxim-Magirus* rearmount aerials was delivered to Philadelphia, Pennsylvania, in 1954. In addition to its own proprietary aerial ladder, introduced in 1947, Maxim installed a number of Magirus aerial ladders on its own custom chassis in both rearmount and midship-mount configurations. Under its licensing arrangement, the Maxim Motor Company also installed Magirus ladders on

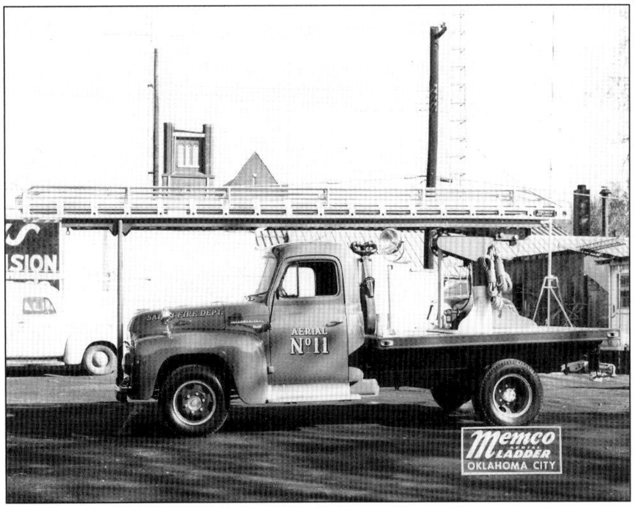

Two makers of light-duty public utility-type aerial ladders — *Memco* and *Sponco* — sold some of their ladders for fire service. Unspecified Salem F. D. cobbled up this practical "baby" aerial with *Memco 55-foot* ladder on an International R-190 platform chassis in the early 1950s. Note the ladder pipe and searchlight behind the cab. *MLC*

American-LaFrance's popular 700 Series service aerials were available with 65-, 75-, 85- or 100-foot steel aerial ladders. Niles, Illinois, got this 1951 *Model 7-65-AJC* aerial with closed canopy-style cab and three-section, 65-foot ladder. Note "zero" overhang of ladder at rear of this apparatus. *WMC*

competitive chassis, too – including Mack and Seagrave, which provided their own bodywork.

After nearly a century, the venerable spring-raised wooden aerial ladder finally went the way of the steamer and buggy whip in the mid-1950s. Oddly enough, it was the major cities – New York, Chicago and Boston – that continued to order spring-raised, wood aerials long after most other departments had abandoned them in favor of hydraulically operated steel aerials. Experienced, well-trained big-city truckies could spot and place a spring-raised wood ladder at an upper-floor window far faster than it took to raise a hydraulic ladder. Spring-raised wood aerials were also easier to maintain and repair.

In 1954, the Chicago Fire Department ordered no fewer than 19 new tractor-drawn, spring-raised wood aerials – twelve from FWD and seven from Peter Pirsch & Sons. Although it had placed its first large order for metal aerials in 1953 – 21 American-LaFrance 700 Series 85-foot tractor-trailer aerials with unique doorless cabs – New York City (FDNY) nonetheless ordered *twenty-five* FWD 75-foot wood aerials – its last – for delivery in 1955. The Chicago Fire Department ordered six more FWD 85-foot wood aerials – the nation's last – in 1956.

One of the U.S. industry's best-known players – the Ward LaFrance Truck Corp. of Elmira Heights, New York – finally entered the aerial ladder truck market in 1955. Ward LaFrance equipped its custom-chassis four-wheel service aerials with 65-, 75- and 85-foot aerial ladders sourced from rival Maxim – which also supplied aerial ladder assemblies to Mack. In addition to its four-wheel types, Maxim by now was also building tractor-drawn aerial ladder trucks with four-section 85- and 100-foot aerial ladders. Maxim's tillered aerials had fixed tiller seats and windshields – a la American-LaFrance.

Also in 1955, Seagrave broadened its aerial offerings with a new lightweight junior aerial ladder mounted on a commercial truck chassis. Aimed at smaller communities with limited budgets, the midship-mounted, three-section 65-foot ladder was available on standard Ford, Chevrolet/GMC, Dodge or International Harvester conventional truck chassis with open or closed cab. Despite its pedigree, Seagrave's commercial aerial was never a strong seller and had been quietly discontinued by the end of the decade.

Slowly but steadily the compact rearmount aerial concept promoted by Maxim with its Magirus connection was gaining favor. More and more fire chiefs began to recognize the inherent advantages of the rearmount configuration – ease of handling in traffic, minimal space required on the street in front of the fire building, and the fact that a rearmount aerial could easily fit into nearly any firehouse. Philadelphia now had two Maxim/Magirus in service. The Milwaukee Fire Department purchased three Mack "B" Magirus 100-foot rearmount aerials with open semi-cabs in 1957. That same eventful year, Green Bay, Wisconsin, placed into service what was then the tallest aerial in North America – a six-section 146-foot Magirus rearmount on a Seagrave 70th Anniversary Series chassis.

The *Four-Wheel-Drive Auto Company* of Clintonville, Wisconsin, had also entered this burgeoning new ladder truck market in 1956. Like rival Maxim, FWD hooked up with an established European aerial ladder manufacturer – Geesink, of Weesp, Holland. The company exhibited its prototype FWD/Geesink aerial, with 85-foot ladder and open semi-cab, at the 1956 IAFC conference in Miami, Florida. Four standard ladder lengths were available – 65, 75, 85 and 107 feet.

Up until 1958, the longest ladder available on a Seagrave four-wheel service aerial was a four-section 85-footer. Seagrave delivered its first four-section 100-foot service aerial to Cheltenham, Pennsylvania, in 1958. Syracuse, New York, and Cleveland Ohio, placed Seagrave 100-foot service aerials into service in 1959. Rival American-LaFrance had offered a 100-foot four-wheel service aerial since 1940.

But American LaFrance's biggest news for 1958 was the introduction of the successor to the company's industry-leading – but now dated – 700/800 Series. Announced in the spring of 1958, the company's all-new *900 Series* pumpers and aerials featured a wide, smooth-faced new cab design with a huge curved windshield and dual head and parking light bezels. The first 900 Series aerial delivered to a fire department was a tractor-drawn 100-footer shipped to Pottsville, Pennsylvania, in late October 1958. The 900 Series aerial family included 75-, 85- and 100-foot four-wheel service aerials with semi-open or canopy-style closed cabs, 85- and 100-foot *Rescue Model* tractor-drawn aerials, and quintuple combinations. That same eventful year, American-LaFrance also entered the *commercial* aerial truck market. The Elmira, New York, firm offered a three-section 65- or 75-foot steel aerial ladder on any commercial chassis preferred by the customer – including the tilt-cab International VCO, GMC/Chevrolet and the new Ford C-Series, as well as on Ford, Chevrolet and IHC conventionals.

In 1959, the Chicago Fire Department purchased four massive Mack/Magirus rearmount aerials – two with 100-foot ladders and two more with six-section 146-foot Magirus ladders. Constructed on special

By the early 1950s, American-LaFrance's comprehensive 700 Series model line had been broadened to include five-way *Quintuple Combinations.* This 1952 *Model 7-75-QJO* with 750 gpm pump and 75-foot steel aerial was built for the U.S. Canal Zone. Note hard suction hose and busy pump-control panel arranged around manual outriggers. *WMC*

Mack offered 65- and 75-foot service aerials on its handsome *L-Series* custom chassis from 1949 through 1953, when the classic *"L"* was replaced by the all-new "B" Series. All were equipped with Maxim ladders. Here's a head-on view of a late-production Mack "L" Series aerial of the early 1950s. *MLC*

Originally introduced in 1940 and nearing the end of its long production life, Mack "L" Series fire apparatus line of early 1950s included quads, service aerials and quints. This heavily laden L-Series quintuple combination with 65-foot Maxim aerial and seven-man canopy cab was built for Detroit suburb of Grosse Pointe Farms. *MLC*

Four-Wheel-Drive Auto Company of Clintonville, Wisconsin, began making tractor-drawn wood aerials again in late 1940s. This 85-foot spring-raised aerial with open-cab FWD tractor was delivered to Paterson, New Jersey, in 1952. Note large equipment compartment at front of trailer and swing-away tiller seat and windshield. *MLC*

Mack B-85 chassis with three-man closed cabs, these impressive black-over-red aerials carried no ground ladders and were usually accompanied on alarms by another truck company.

Also in 1959, the Seagrave Corporation finally introduced its first *cab-forward* custom-chassis fire apparatus – twelve years after rival American-LaFrance pioneered this industry-leading configuration. Before the year was out, FWD and Howe had also introduced cab-forward designs of their own. Seagrave's first "K" Series cab-forward aerials were a 75-foot service aerial delivered to Muncie, Indiana, and a pair of 100-foot tractor-drawn aerials for Tucson, Arizona.

The Firefighting *"Snorkel"* Surfaces

"Coming events," it is said, *"cast their shadow before."* Thus it was that a short – but oddly prophetic – news item was published on the *News of the Manufacturers* page in the December 1954 issue of *Fire Engineering.*

Under the headline of *"Mechanical Giraffe Aids Firemen,"* the item concerned the novel use of a *"Pitman Giraffe"* public-utility-type aerial platform at a recent fire in New Westminster, British Columbia. Two hose lines aimed from the rig's bucket helped quell the blaze in a downtown department store.

"This novel unit, although not built expressly for fire suppression, provides a stable platform for fighting fire in upper floors. The versatile boom will rotate 360 degrees...the upper boom moves up and down in a 170-degree arc while the lower boom moves in a 100-degree arc. The device is controlled by the operators on the platform," the article noted.

Originally designed for picking fruit in orchards, the *Giraffe* was invented in the early 1950s by Theodore T. Trump, of Oliver, British Columbia. Pitman Mfg. of Grandview, Missouri, obtained rights to make the *Giraffe* for public utilities. When the elevating platform he had designed was later adapted for firefighting, Ted Trump began making two- and three-boom aerial platforms in his plant in Oliver, a small town in British Columbia's interior.

Fast-forward a few years. On a stroll back to his office after lunch, Chicago Fire Commissioner Robert Quinn reportedly stopped to watch a park tree-trimming crew at work. Quinn was amazed at the mobility the men in the "cherry-picker" truck's bucket had and wondered why Chicago firefighters couldn't have the same kind of mobility. In 1957, Ray Schuster, the Chicago area representative for Pitman Mfg. of St. Louis, Missouri – makers of the *Giraffe*– ordered an A.B. Chance 50-foot articulating boom on a Chevrolet 354 truck chassis. Equipment cabinets were added by Pierce – known then as the *Pierce Auto Body Works* of Appleton, Wisconsin.

Mr. Schuster planned to demonstrate this unit, dubbed the *PLC* for *Pierce Line Construction,* to electric utility companies for power-washing insulators on power transmission lines. But following a convincing demonstration to Chicago Fire Department officials, the unit was immediately sent back to Pitman for further modification and soon there-

Largest and most impressive member of the extremely successful American-LaFrance 700 Series family was the tillered *Rescue Model* tractor-trailer aerial. Most of these long, articulated aerials with fixed tiller seat and windshield had 85- or 100-foot ladders, but a few were built with 75-foot aerials. This classic 100 footer with open semi-cab was delivered to Des Moines, Iowa, in 1953. *WMC*

Boston Fire Department received several of these Pirsch model 41-D 65-foot service aerials with open semi-cabs in 1952-53. Note the scaling ladder mounted over the rear fender and the Sterling siren mounted on the rig's nose. *MLC*

Maxim Motor Company of Middleborough, Massachusetts, revived moribund rearmount aerial concept in U.S. in 1952 with acquisition of North American distribution rights for German-made *Magirus* aerial ladder. Philadelphia, Pennsylvania, purchased several of these short-wheelbase Maxim-Magirus rearmounts aerials starting in 1954. Note full-width bench seat under mechanically raised, four-section, 100-foot ladder. *MLC*

Maxim began building tractor-drawn, tillered aerial ladder trucks in the early 1950s. This 85 footer with open semi-cab and double-bar front bumper went to Philadelphia, Pennsylvania. Note stanchion-mounted *Buckeye Roto-Ray* warning light on cowl. *WMC*

after placed into actual fire service. The nation's first firefighting aerial platform, this rather crude-looking apparatus proved its worth at a lumberyard fire a few weeks later. Drenched with spray bouncing off roofs and walls, the Chicago firemen in the basket jokingly nicknamed their experimental rig *Quinn's Snorkel.* The moniker stuck, and the *Snorkel* entered the fire service lexicon – and history.

Following the spectacularly successful 1958 experiment, Chicago ordered two more *Snorkels* on Ford conventional truck chassis with 65- and 75-foot *Hi-Ranger* articulating booms made by *Mobile Aerial Towers* of Fort Wayne, Indiana.

The elevating platform made its formal fire service debut at the 1959 annual convention of the International Association of Fire Chiefs held in Grand Rapids, Michigan, where Pitman exhibited a 65-foot Snorkel on Ford tilt-cab chassis. Following a demonstration tour, this unit was sold to Elwood, Indiana. It's still there today, remounted on a '77 Ford chassis.

The 1959 IAFC Convention issue of *Fire Engineering* featured ads from half a dozen aerial platform makers eager to get in on this promising new trend. In addition to Pitman Mfg. and Hi-Ranger, four other companies advertised aerial platforms: the *Strato-Tower Division* of Young Spring & Wire, Elkhart, Indiana; the *J.H. Holan Corp.* of Cleveland, Ohio,

Maxim 75-foot service aerial in action at a multiple-alarm fire in Indianapolis, Indiana. Ladder pipe operator at tip of ladder signals change of ladder direction to operator on turntable. Note also the old Stutz high-pressure rig at right. *MLC*

makers of the *Holan Aerial Elbow;* the Truck Equipment Co. of Denver, Colorado, makers of the *Truco Aerial Lift;* and the Peters Co. of Portland, Oregon. Which offered a vertical-rise platform called the *Industrial Monkey*!

By the end of the year, two Chicago area fire departments – Bedford Park and Loves Park – had Snorkels in service. Virtually overnight, the firefighting aerial platform had arrived. The "aerial revolution" was underway.

Seagrave introduced all-new *70th Anniversary Series* in mid-1951. The company's first postwar design, *Anniversary Series'* most distinctive feature was nose-mounted siren. Among early deliveries was this 65-foot service aerial with open semi-cab for St. Paul, Minnesota. *MLC*

Trio of *Pirsch Senior* 100-foot tractor-drawn aerials delivered to Rochester, New York, in 1953 make impressive aerial pyramid shortly after delivery. Rochester Fire Department got five Pirsch 65- and 75-foot service aerials in 1946-47. *MLC*

Detroit Fire Department purchased several of these Seagrave 65-foot service aerials with rear-entrance canopy-style cabs and booster equipment in the early 1950s. Note side-mounted booster hose reel and short "Dutch town" ladder above life net compartment. *WMC*

Loyalty begins at home! Peter Pirsch & Sons Company's home city of Kenosha, Wisconsin, took delivery of this *Model 41-E Senior Tractor-Drawn Aerial* with three-section, 100-foot ladder and closed three-man cab in 1954. Note grille-mounted bell, siren and Mars warning light. *WMC*

Chicago Fire Department received twelve of these FWD *Model FA-85K* tractor-drawn aerials with closed cabs and 85-foot spring-raised aerials in 1953-54, and six more — its *last* wood aerials — in 1956. Note C.F.D.'s distinctive black-painted upper cab, football-shaped Mars *DX* light on roof and front-mounted bell. *MLC*

Despite fire service shift to hydraulically operated metal aerial ladders, New York, Boston and Chicago still preferred old-style spring-raised, wood aerials. Experienced, well-trained crews could spot and raise one of these wood "sticks" far faster than a metal one. Chicago Fire Department received six Pirsch *Model 41-E tractor*-drawn 85-foot aerials in 1954. New York and Chicago got their last wood aerials in 1955-56. *MLC*

Two tractor-drawn Seagrave aerial trucks at extra-alarm fire on narrow residential street in Boston, Massachusetts. Ladder 13's rig in foreground is a 1958 Seagrave 100 footer. Ladder 17's rig parked further down the street is a `56 Seagrave. *BFD Photo/Bill Noonan*

Ward LaFrance Truck Corp. of Elmira Heights, New York, entered burgeoning U.S. aerial ladder truck market in 1955. Built on the company's distinctive custom chassis, new Ward LaFrance service aerials were fitted with Maxim 65-, 75- or 85-foot steel aerials. This handsome 1956 Ward LaFrance Quint with Maxim 75-foot ladder went to DeKalb, Illinois. *WMC*

Several smaller U.S. fire apparatus manufacturers, including *Howe, Boyer* and *American Fire Apparatus*, were building aerial trucks and quints with *Memco* utility type aerial ladders. Madison, Indiana, got this *Boyer* quint built by *Universal Fire Apparatus Corp.* of Logansport, Indiana, on a 1955 Diamond-T chassis with electrically operated *Memco* 65-foot aerial ladder. *Dave Stewardson*

Mack replaced prewar design "L" Series with all-new "B" Series conventionals in 1953. Berwyn, Illinois, took delivery of this good-looking "B" Series 75-foot aerial with open semi-cab and booster equipment in 1955. *WMC*

Unlike the earlier L/LS Series, Mack's all-new *"B" Model* fire apparatus line included both straight-frame service and tractor-drawn aerial ladder trucks. Most had Maxim aerials, but some were also delivered with Magirus ladders. Here's a factory-fresh B85F tractor-trailer aerial with open semi-cab and Maxim 85-foot ladder photographed in the plant yard in Allentown, Pennsylvania. *MLC*

New York City bought more American-LaFrance 700 Series fire apparatus than any other city — most of them 85-foot tractor-drawn metal aerials like this one delivered in 1956. But the 21 delivered to the FDNY in 1953 and 11 more received in 1956 differed from all others in that they had *no* cab doors. The `56s also had extra-high windshields. *MLC*

Until the late 1950s, the longest aerial ladders made in the U.S. were 100 footers introduced in the 1930s. In 1957, Green Bay, Wisconsin, ordered this huge Seagrave 70th Anniversary Series aerial with rearmounted *Magirus* 146-foot ladder. Two years later Chicago bought two Macks with 146-foot Magirus ladders. New York City (FDNY) joined the Mack/Marius high-ladder club in 1961. *Dan G. Martin Photo*

In 1955, Seagrave introduced an all-new, lightweight aerial ladder designed specifically for mounting on commercial truck chassis. Seagrave's 65-foot, three-section *commercial aerial* was considerably less expensive than the company's custom-chassis 70th Anniversary Series aerials. This one, on 1959 Ford F-Series chassis with cut-down, semi-open cab, was delivered to Adams, Massachusetts. *MLC*

Carefully spaced according to strict safety guidelines for aerial ladder operation, five Syracuse, New York, firefighters climb 100-foot aerial ladder during acceptance test of new 1957 American-LaFrance 700 Series service aerial at Syracuse Fire Department Drill Yard. Note nose jack under front bumper. *Paul Norman Photo*

By the late 1950s, Seagrave was also delivering some commercial-chassis aerial trucks equipped with the company's *standard* aerial ladder — the same one used on the company's custom chassis. Brockport, New York, got this compact 75-foot straight aerial on 1959 International tilt-cab chassis, seen here in action at canning factory fire in 1977. *WMC*

In 1958, American-LaFrance — for the first time — began mounting its standard 65- and 75-foot aerial ladders on commercial truck chassis. Endwell, New York, ordered this compact, well-equipped quint with 75-foot ladder and 750 gpm pump on International VCO tilt-cab chassis. Note the nose-mounted Federal "Q" siren. *WMC*

Pirsch Senior tractor-drawn aerial of the late 1950s. Note new box-style front fenders, handy platform for the bell on the right and *B&M* siren on the left, and *Mars* light atop the grille. Jackson, Mississippi, received this imposing *Model 41-E* with three-section, 100-foot aerial in March 1958. *MLC*

Here's the truck that sparked a revolution in aerial firefighting in America. Chicago Fire Department placed this GMC flatbed with public utility-type *Pitman Snorkel* into service as an experiment in 1958. Commissioner Robert Quinn's *Snorkel* proved so successful that C.F.D. ordered three more the following year — and the rest is history. *Bill Friedrich Photo*

Peter Pirsch & Sons continued to offer its lightweight *Pirsch Junior* 55-foot aerial ladder on commercial *or* Pirsch Custom chassis. The late John F. Sytsma — an inveterate parade picture-taker — found this 1950s vintage *Pirsch Junior* at a firemen's parade in Canton Township, Ohio. *JFS/WMC*

Turntable view of a Seagrave aerial ladder of the 1950s. Note the heavy crossmember across the top of the ladder — which firefighters had to clamber over on their way up. The cable drum that extended the two fly sections is visible at the bottom of the photo. *MLC*

In 1957, the *Four Wheel Drive Auto Co.* of Clintonville, Wisconsin, became the U.S. distributor for *Geesink* aerials imported from Holland. Harvey, Illinois, received this FWD rearmount aerial with four-section, 107-foot *Geesink* ladder in 1959. *MLC*

In 1956, Mack Trucks acquired Sidney, Ohio, bus maker C.D. Beck — which also built Ahrens-Fox fire apparatus. Mack appropriated excellent Ahrens-Fox cab-forward fire apparatus design — which became Mack's hugely successful C-Series. New York City purchased a large number of Mack C-Model pumpers and aerials in 1958-59 — including this tractor-drawn aerial with Maxim 85-foot ladder. *MLC*

In 1958, Seagrave for the first time began offering a four-section, 100-foot aerial ladder on its straight-frame service aerial chassis. Up to now, all Seagrave 100-foot aerials were of the tractor-drawn type with long three-section ladders. Parma, Ohio, got one of the first of these 70th Anniversary Series 100-foot service aerials in 1958. *JFS/WMC*

Spring-raised wood and hydraulic steel aerials at work as water towers at commercial building fire in Windsor, Ontario, in 1963. In foreground is a 1936 American-LaFrance 85-foot tractor-drawn aerial. The metal aerial is a 1952 Bickle-Seagrave 75 footer. *WMC*

Born in Chicago just two years earlier, the elevating platform, or *Snorkel*, spread rapidly across Windy City suburbs, and across the nation. One of the first sold outside Chicago was 50-foot *Pitman Snorkel* on tandem-axle GMC chassis with Pierce bodywork delivered to Bedford Park, Illinois. *MLC*

CHAPTER SEVEN: THE 1960s

The 1960s witnessed the swift, widespread adoption of the articulated elevating platform by the American fire service, the introduction at mid-decade of the competing *telescopic*-type aerial tower, and – late in the decade – belated acceptance of the *rearmount* aerial ladder truck.

Following its sudden appearance in 1958, the elevating platform spread through the North American fire service like a prairie grass fire – initially in the Chicago area, then across the U.S. and Canada. Led by Chicago, major cities across the U.S. began to retire their ancient water towers, replacing them with this remarkably versatile new kind of aerial master stream device.

By the early 1960s, three different firms were marketing articulating-boom aerial firefighting devices using the *Snorkel* name. In 1958, as the elevating platform evolved from novelty to a potent firefighting weapon, the Pitman Mfg. Co. formed a separate division to make and market the *Pitman Snorkel.* The *Snorkel Fire Equipment Company* moved into its very own plant in Grandview, Missouri. Mobile Aer-

Steger, Illinois, was another early *Snorkel* purchaser. Here Steger's Pierce-bodied 1960 FWD with 50-foot *Pitman Snorkel* is in action at a major fire in that community. Note outriggers at base of twin-boom articulating platform. *MLC*

ial Towers of Fort Wayne, Indiana – makers of the open-lattice *Hi-Ranger* aerial platform, marketed its product as the *Hi-Ranger Snorkel.* And in 1960, Ted Trump – inventor of the original *Giraffe* – formed *Trump Ltd.* in Oliver, British Columbia, to make and market the *Trump Snorkel.* Fredericton, New Brunswick, placed Canada's first elevating platform – a 65-foot Trump Snorkel on Chevrolet conventional truck chassis – into service in 1960.

The first Snorkels were dedicated, single-purpose aerial master stream devices, usually called in to knock down big fires, or for rescue duty. Within no time, however, the Snorkel proved itself amazingly adaptable to a variety of fire apparatus configurations. In 1959 Elko, Nevada, got the first Snorkel-equipped triple combination pumper. Built by Crown Coach of Los Angeles, the tilt-cab International VCO pumper was fitted with a 750 gpm pump and 50-foot Snorkel. In 1961, Memphis, Tennessee, ordered the first Snorkel with a full complement of ground ladders, qualifying it to respond to alarms as a regular truck company. Several fire departments retrofitted older pumpers with 50-foot Snorkels, vastly enhancing their usefulness.

While the newfangled Snorkel seemed to be getting all the attention as the `60s began, the conventional steel aerial ladder still reigned supreme. In 1960, Seagrave delivered its first tractor-drawn aerials with compact four-section ladders and fixed tiller seat/windshield. Philadelphia, Pennsylvania, got three 70th Anniversary Series 100-foot aerials of this style in 1960-61. The Columbus, Ohio, plant was also building fixed-tiller aerials of this type in its new custom cab-forward "K" Series.

In 1961, New York City (FDNY) joined the "high-ladder club" with the purchase of two Mack C-Series aerials equipped with rear-mounted six-section 146-foot Magirus aerials. At about the same time, Pirsch and Maxim introduced their own custom cab-forward chassis, which soon proved ideal for straight-frame and tractor-drawn aerials, matching those made by American LaFrance and Seagrave.

Seagrave's bread-and-butter custom was still the siren-in-the-nose *70th Anniversary Series* introduced a decade earlier. Covington, Kentucky, got this fully loaded Seagrave *Model 900-AB Quintuple Combination* with 1000 gpm pump, 300-gallon booster tank and compact four-section 75-foot aerial. *John Denney Photo*

Seagrave finally introduced its first cab-forward fire apparatus in 1959 — twelve years after rival American-LaFrance. Its new *"K" Series* was well received and immediately far outsold the 70th Anniversary Series conventional. Minerva, Ohio, got this *Model KA* quint with 750 gpm pump and 75-foot aerial ladder in 1960. *John F. Sytsma*

Through the 1950s two Oklahoma-based makers of public utility-type aerial ladders – *Memco* and *Sponco* – offered lightweight 50- to 65-foot aerial ladder assemblies for mounting on commercial chassis fire apparatus. American Fire Apparatus and Howe, among others, delivered pumpers equipped with these electrically operated two- and three-section, low-handrail ladders. In the early 1960s, *Grove Mfg. Ltd.,* a well-known maker of cranes and industrial ladders located in Shady Grove, Pennsylvania, began making its heavy-duty ladders available on an aftermarket basis to independent fire apparatus builders. American and Howe were early customers for Grove aerials, which were made in three-section 85-foot and four-section 100-foot sizes. The distinctive Grove ladder featured a deep-section bed ladder with tapered tip.

But perhaps the most *startling* news of the early 1960s was the delivery of the industry's first *jet-powered* fire apparatus. American-LaFrance exhibited America's first turbine-powered pumper at the 1960 IAFC convention in Rochester, New York. Powered by a 325-horsepower Boeing industrial turbine, the 900 Series *Turbo-Chief* pumper was sold to San Francisco. A year later ALF delivered the nation's first jet-engined aerial ladder truck to Boeing's home city of Seattle, Washington. The 900 Series *Turbo-Chief* 100-foot tractor-drawn aerial with canopy cab tractor and fully enclosed ladder bed was also powered by a Boeing turbine. A second ALF *Turbo-Chief* turbine pumper later went to Mount Vernon, Virginia. Alas, however, this high-profile experiment was less than a resounding success. Among other problems, the whining turbo's lack of engine braking power meant repeated, costly brake jobs. After just a few years of service, all three – including Seattle's aerial – were quietly converted to conventional gasoline power.

As the decade unfolded, the *Pitman Snorkel* and *Hi-Ranger* dominated the booming aerial platform market. *Snorkel* was by far the largest player, with 55-, 65-, 75- and 85-foot articulating boom platforms available on a wide range of custom and commercial chassis. During the 1960s Pierce Mfg. of Appleton, Wisconsin, built hundreds of big snorkel bodies for various fire apparatus manufacturers, including FWD, Pirsch, Mack, Maxim and Crown. Try as they might, the industry's two largest players – American-LaFrance and Seagrave – could no longer ignore this increasingly important new market segment.

Seattle, Washington, got the only *jet-powered* aerial ladder truck built in America. Powered by a 325-horsepower Boeing turbine, 900 Series 100-foot tractor-drawn *Turbo-Chief* with enclosed ladder bed was delivered in 1961. Less than a success, within a few years it was quietly converted to conventional gasoline power. *MLC*

Crown Firecoach of Los Angeles began building elevating platforms on custom and commercial chassis in the early 1960s. Downey, California, got this *Crown Firecoach* with open semi-cab and 65-foot *Pitman Snorkel* in 1961. *WMC*

American-LaFrance was the first to respond. The Elmira, New York, firm unveiled its first elevating platform at the 1962 IAFC convention held in Toronto, Canada. The company's 900 Series *Aero-Chief* was equipped with a 70-foot Strato-Tower articulating boom made by the Daybrook Division of Young Spring and Wire. A unique feature of the prototype *Aero-Chief* was its hinged, two-piece windshield which could be opened at its center to allow the boom to be lowered in its travel position to clear underpasses or other low-overhead obstacles. *Aero-Chief* platform deliveries began the following year.

At about the same time, Seagrave announced its *Cemco-Eagle* elevating platform, to be offered in twin-boom 65- and 85-foot versions. The Seagrave Corporation – acquired by the Four-Wheel-Drive Auto Company (FWD) of Clintonville, Wisconsin, in mid-1963 – delivered its first aerial platforms this year.

Nineteen-sixty-three was perhaps the most eventful single year in this frenetic decade. In mid-1963, American-LaFrance delivered its first production *Aero-Chief* elevating platform to Juneau, Alaska. Where the 1962 prototype (and another on 1963 International VCO tilt-cab chassis) were fitted with Daybrook articulated booms, the one shipped to Juneau had a 70-foot, two-section boom of ALF's own design and manufacture. The *Aero-Chief's* upper boom nested neatly into the hollowed-out upper end of the base section permitting a lower travel height. After a demonstration tour, the 1962 *Aero-Chief* prototype was sold to Berkeley, California.

Rival Seagrave delivered its first elevating platforms in 1963. The only one ever constructed on the company's famed siren-in-the-nose 70th Anniversary series custom conventional chassis went to Nashville, Tennessee. The two-section 70-foot *Eagle* boom was of unique semi-cylindrical design. Three more – all on the company's K-Series custom cab-forward chassis – had 90-foot, three-section Daybrook booms. These massive *Seagrave Eagle* aerial platforms went to Kentland, Maryland, San Bernardino, California, and Hampton, Virginia. Also in 1963, Ward LaFrance began deliveries of *Hi-Ranger* elevating platforms on its *Ambassador* custom cab-forward chassis.

The next significant advance in aerial platform evolution wasn't long in coming. Founded in 1890, *C.H. Sutphen & Sons* later became a dealer for the Prospect Fire Engine Co. By the 1950s the Columbus, Ohio, fire equipment firm was a regional distributor for Pirsch and American Fire Apparatus. Sutphen was also making some commercial chassis apparatus under its own name. In the early 1960s, Sutphen designed an entirely new *kind* of elevating platform. Instead of the usual rear-mounted articulated boom, the *Sutphen Aerial Tower* had a midship-mounted, *telescopic*, three-section aluminum-alloy box-type boom of riveted open-lattice construction, with integral prepiped waterway. A crew basket with monitor nozzle was attached to the upper end of the

New York City joined the exclusive "High-Ladder Club" in 1961 with delivery of two Mack C-Series rearmount aerials with 146-foot Magirus ladders. After a decade of service, the massive six-section Magirus ladders were replaced with more conventional Grove 100-foot ladders. *MLC*

Long Island fire companies were famed for lavishly appointed fire apparatus. Seagrave delivered this tractor-drawn *KT-100* aerial with what is believed to be the nation's first fully enclosed tiller cab to Freeport, New York, in 1961. Seagrave made its first aerials with a compact four-section, 100-foot ladder and fixed tiller seat circa 1960. *WMC*

Mack continued to build popular *B-Series* conventional fire apparatus alongside its new cab-forward *C-Series*. Mack aerials were available with either Maxim or Magirus ladders. Gulfport, Mississippi, ordered this *B-Model* Mack with open semi-cab and 100-foot, four-section *Magirus* aerial ladder. *MLC*

fly section via a patented yoke that kept the platform level at all times.

The prototype 65-foot *Sutphen Aerial Tower* was mounted on the ubiquitous Ford C-Series tilt-cab chassis. The industry's first *aerial tower* was exhibited at various chiefs' conventions that year. That pioneering first *Sutphen Tower* was later sold to the city of Norwalk, Ohio. This extremely successful innovation from one of the smaller regional manufacturers was a real game-changer. More than a thousand *Sutphen Aerial Towers* have been delivered in the four decades since and to this day remain this nationally respected company's signature product

Mack introduced the industry's *second* telescopic aerial tower in 1964. Mounted on Mack's C-Series custom cab-forward chassis, the *Mack Aerialscope* had a midship-mounted 75-foot four-section *Truco/Eaton* boom of solid welded steel box construction with a crew platform and prepiped monitor nozzle at its tip. Six hydraulically operated outriggers steadied the bulky-looking apparatus when the boom was raised. The *Mack Aerialscope* almost immediately became the aerial weapon of choice for the nation's largest and busiest fire department – New York City. The FDNY placed its first Mack Aerialscope into service in 1964. Another one of these rugged *Tower Ladders* – as the FDNY dubbed them – was purchased two years later, followed by five more in 1969. Several hundred more – the majority of them on Mack chassis with Baker booms – have been delivered since and 75- and 95-foot *Aerialscopes* on Seagrave custom chassis remain the backbone of the FDNY's aerial firefighting fleet to this day.

Enter The Rearmount

In the mid-1920s, A German manufacturer – *Magirus* – introduced the compact, short-wheelbase, rear-mounted aerial ladder truck concept to the North American fire service. The standard aerial configuration everywhere else in the world *except* North America, the "continental prototype" rearmount aerial simply never caught on with U.S. and Canadian fire departments. Fewer than a dozen were sold before *American Magirus* closed up shop at the depths of the Depression.

Over the three succeeding decades, several manufacturers – including such prominent names as American-LaFrance and Ahrens-Fox – toyed with the rearmount ladder configuration, but these design proposals never made it off the drawing board. Seagrave actually delivered two rearmount aerials, both for industrial customers. The first, built by Yankee on a 4X6 International KB chassis in the late 1940s had a Seagrave 75-foot aerial ladder with a one-man bucket at its tip. The second, with 65-foot rearmount ladder on a Chevrolet truck chassis, was made in 1955.

In August of 1964 Seagrave announced its *Rear Admiral* – a compact cab-forward aerial truck with the ladder's turntable at the rear of the chassis, with the four-section 100-foot aerial extending forward over the open semi-cab. This design reduced overall length of the apparatus by a third compared with the conventional midship-mounted service and tractor-drawn aerial ladder trucks being built at the time. What's more, the *Rear Admiral* was also available with four-wheel steering. Seagrave claimed the rearmount layout required less room in the fire station and made the 224-inch wheelbase unit easier to maneuver in traffic or at the fire scene. A prototype was constructed on a cab-forward FWD chassis. After extensive testing and demonstrations, the first production Seagrave *Rear Admirals* were delivered in 1967. Among the first customers was the Chicago Fire Department, which purchased many more in later years.

American-LaFrance countered with its 900 Series *Ladder-Chief* in 1968 – the first of which was delivered to Gary, Indiana. Within a few years every one of the major U.S. fire apparatus manufacturers, including Pirsch, Maxim, Mack, Crown, Howe and Hahn, had added rearmount aerials to their product lines.

Also in 1968, *Snorkel Fire Equipment* introduced an entirely new kind of aerial device. The *Squrt* was essentially a 54-foot articulating water tower with a remotely controlled master stream nozzle at its tip instead of a basket. Resembling a scaled-down Snorkel, the *Squrt* – with a 45-foot horizontal reach – was designed for easy installation on a new or any older model pumper or hose wagon. The Squrt's pedestal required just 18 x 28 inches of the truck's hose bed. Following its introduction at the 1968 Fire Department Instructors Conference (IAFC) and an extensive demonstration tour, the pilot model *Squrt* – on a Ford C-Series chassis with compact squad-type rear body – was sold to Indianapolis, Indiana. The first pumper-mounted Squrt went to Michigan City, Indiana.

Up to now, the tallest elevating platforms and aerial towers available had a maximum working height

American LaFrance entered the burgeoning elevating-platform market in 1962. The company's first two elevating platforms were equipped with *Daybrook* booms sourced from Young Spring & Wire. Prototype 900 Series *Aero-Chief* with 70-foot boom seen here was eventually sold to Berkeley, California. *WMC*

Mack's 1956 acquisition of C.D. Beck included Ahrens-Fox's promising new cab-forward fire apparatus design — which became Mack's extremely successful *"C" Series.* With a choice of Maxim or Magirus ladders, the new C-Series was available in both four-wheel service and tractor-drawn types. Glen Ridge, New Jersey, got this C-Series Mack with Maxim 75-foot aerial. *MLC*

Young Fire Equipment Corp. of Buffalo, New York, delivered a number of *Snorkel* elevating platforms on Ford heavy-duty truck chassis. Westhampton Beach, Long Island, New York, got this Young-built 65-foot *Pitman Snorkel* on Ford F-Series chassis with cut-down open semi-cab in 1965. *Shaun P. Ryan Photo*

In a fireground scene that would become increasingly familiar across North America, a conventional aerial ladder truck (900 Series ALF) and an 85-foot *Pitman Snorkel* work side-by-side at supermarket fire in Albany, New York, in 1962. Note how articulating boom reaches up and over power lines, walls, etc. to knock down fire from directly above. *MLC*

Snorkel elevating platform at left, and American-LaFrance 700 Series service aerial in action at a school fire in 1970s. Elevating platform gave firefighters unprecedented aerial mobility without having to clamber up and down a ladder. *WMC*

Along with pioneering *Pitman,* a Canadian manufacturer, *Trump Ltd.* of Oliver, British Columbia, delivered many early elevating platforms — including the fledgling industry's first three-section 90 footers. Trump also used the *Snorkel* name, as seen on grille of this three-boom tower for Winnipeg, Manitoba. Note massive A-frame outriggers behind the GMC three-man cab. *WMC*

Then a relatively small regional fire equipment distributor, *Sutphen Corp.* of Columbus, Ohio, introduced a whole, new dimension in aerial firefighting with the groundbreaking *Sutphen Aerial Tower* in 1963. Three-section, 65-foot telescopic boom of bolted, lattice construction was mounted on Ford tilt-cab chassis. Prototype *Sutphen Tower* was later sold to Norwalk, Ohio. *WMC*

Seagrave also entered the booming aerial platform market in the early 1960s. One of Seagrave's first aerial elevating platforms was one-of-a-kind *Eagle* 70 footer on 70th Anniversary Series chassis built for Nashville, Tennessee, in 1963. Note tubular twin-boom construction. *Joel Woods/Shaun Ryan Collection*

of 85 feet. But in 1969, the Calavar Corporation of Santa Fe Springs, California, announced a giant new dimension in aerial firefighting. This company's monstrous *Calavar Firebird* aerial platform towered 125 feet above the street. In general appearance the *Firebird* looked like an oversized version of a standard dual-boom articulating elevating platform. But what set the mighty *'Bird* apart from all others was the fact that *both* of its booms – upper and lower – were telescopic, virtually doubling each arm's length. With a working height of an unprecedented 125 feet, the Firebird also boasted a straight-out *horizontal* reach of 78 feet!

Dual hydraulic lifting cylinders held the base column and upper boom rigidly in place during operation. Watching one of these humongous aerial towers go into action at a fire for the first time was an unforgettable experience. Carefully spotted in front of the fire building, the *Firebird* sprouted two pairs of huge H-shaped hydraulic outriggers, which spanned nearly the entire width of the street. Enclosed, hydraulically operated ground ladder boxes on each side of the truck dropped down 45 degrees to permit the massive base column to rotate 360 degrees. Then, with its tall single nose jack down, the entire ten-ton, tandem-axle monster hoisted itself up 18 inches off the pavement. Quite a show...

Philadelphia, Pennsylvania, took delivery of the first *Calavar Firebird 125* aerial platform – the tallest aerial platform in America at the time – in 1969. Built on a special Seagrave custom six-wheel chassis with five-man canopy style closed cab, the mighty *Firebird* was assigned to the P.F.D.'s Ladder Co. No. 2 where it served for nearly a decade.

As the 1960s wound down, articulated and telescopic-type aerial platforms had replaced or supplemented conventional aerial ladder trucks at hundreds of fire departments large and small from coast to coast.

The other less-visible milestone development in fire apparatus of the 1960s was the dramatic shift from gasoline to *diesel* power. By the early 1960s, American LaFrance and Seagrave had discontinued their venerable V-12 engines, offering instead large inline gasoline engines sourced from Continental, Waukesha and Hall-Scott. Championed by Mack, the "diesel revolution" began in earnest in the mid-1960s. The simpler, more fuel-efficient diesel proved especially suitable for heavy apparatus like aerial ladder trucks and elevating platforms. Besides Mack – which had delivered its first diesel-powered fire apparatus in 1960 – other OEM diesel engine suppliers included the Detroit Diesel Division of General Motors, Cummins and Caterpillar.

Nashville's 1963 Seagrave *Eagle Rescue Unit* works with Seagrave and Pirsch conventional aerials at three-alarm furniture store fire on Broadway in 1970s. *Charles W. Vick*

Seagrave delivered two of these big aerial platforms with *Trump* 90-foot, three-section booms in 1963 — this one to San Bernardino, California, and a similar unit to Kentland, Maryland. Note midship location of turntable, low-mounted A-frame outriggers, tandem rear axles and enclosed ladder beds. *WMC*

In addition to proprietary *Strato-Tower* and *Eagle* elevating platforms, Seagrave Corp. also built some custom-chassis aerial platforms with *Pitman Snorkels.* This one, with 1000 gpm pump and 65-foot *Pitman Snorkel,* went to the Detroit suburb of Grosse Pointe Woods, Michigan in 1964. *WMC*

New York City's first elevating platform was the prototype 1964 *Mack Aerialscope.* FDNY has purchased hundreds since. Designated *Tower Ladders* by FDNY, Mack 75-foot Aerialscopes are in the thick of action at every multiple-alarm blaze in Gotham. Here, a first-generation, 1960s-vintage Mack Aerialscope rises to the occasion. *MLC*

Midship-mount *telescopic* aerial towers debuted in the mid-1960s to challenge articulated types that originally sparked the elevating platform revolution. *Mack Trucks* announced its milestone *Aerialscope* in 1964. Early *Mack Aerialscopes* had Eaton or Truco booms. The first *Mack Aerialscope* — on C-Series custom chassis with three-section, 75-foot boom — went to New York City (FDNY). *WMC*

Among *Snorkel Fire Equipment's* most popular products was a 50-foot Snorkel designed for mounting on any new — or older model — pumper. Vandergrift, Pennsylvania, ordered this versatile pumper/snorkel on 1964 Ford C-950 chassis with 1000 gpm pump and Pirsch bodywork. *WMC*

Since introduction of the cab-forward "K" Series in 1959, Seagrave was delivering fewer and fewer *70th Anniversary Series* conventionals. Most (but not all) late-production 70th Anniversary Series aerials had four-section ladders and fixed tiller seats. This 100 footer with canopy cab tractor went to Seagrave's hometown of Columbus, Ohio in 1965. *MLC*

A significant number of early elevating platforms combined components of three companies. FWD supplied long-wheelbase chassis with Cincinnati canopy cab, Pierce Mfg. of Appleton, Wisconsin, built most of the big compartmented bodies, and the *Pitman Snorkel* boom came from Snorkel Fire Equipment Co. Detroit got this standard-issue FWD/Pierce/Pitman 85 footer in 1965. *MLC*

Many older rigs were retrofitted with *TeleSqurt* aerial devices. Carmel, Indiana's 1965 Howe front-mount pumper/tanker on International CO-8190 chassis was later fitted with a 55-foot *Telesqurt* combination aerial ladder/water tower. *WMC*

In Canada, *King-Seagrave Ltd.* of Woodstock, Ontario, was delivering aerial platforms with *Daybrook* booms made by Young Spring & Wire. Toronto Fire Department's first elevating platform was this imposing 85 footer on 1965 Ford F-Series chassis. Hamilton, Ontario, got an identical unit on 1964 Mercury chassis. *Alan M. Craig Photo*

Pennsylvania crane maker *Grove Mfg.* began making fire department-type aerial ladders in 1959. The first conspicuously successful independent aerial ladder maker in the U.S., Grove emerged as an increasingly popular OEM manufacturer in the 1960s. A typical Grove installation is this 100-foot, four-section ladder on Ford C-Series quint built by American Fire Apparatus Co. for Ann Arbor, Michigan. *WMC*

Prototypical American LaFrance *900 Series* 100-foot *Service Aerial* of the mid-1960s. ALF aerials still had manually operated, mechanical outriggers like those visible in this photo. *MLC*

The fourth *Sutphen Aerial Tower* delivered was this 85 footer on 1966 International CO-8190 chassis with Cincinnati cab built for Oak Brook, Illinois. Note massive hoist construction and low mounting of crew bucket with dual side-mounted turret nozzles. Most later Sutphen Towers had tandem rear axles. *WMC*

Very few *Mack Aerialscopes* were built with open cabs. St. Louis, Missouri, got this 75 footer on C-Series custom chassis in 1966. Note vertical stabilizers forward of cab doors and drop-down stabilizers behind mid-mounted turntable. *WMC*

Along with rival Pitman, *Mobile Aerial Towers* of Fort Wayne, Indiana, maker of the open-lattice *Hi-Ranger,* was one of the true pioneers of the firefighting elevating platform. While most early *Hi-Rangers* were on Ford chassis, *Mobile Aerial Towers* later developed a relationship with Ward LaFrance. One of many *Hi-Ranger* booms mounted on square-rigged Ward LaFrance *Ambassador* custom chassis, this 75 footer went to Cheektowaga, New York. *MLC*

Early *Sutphen Aerial Towers* were on heavy-duty commercial truck chassis — mostly tilt-cab Fords and GMCs. Cleveland, Ohio, got this 1968 Sutphen/GMC with canopy cab extension and 85-foot boom. Note ground ladders stacked above rear fender compartment and coupling for tower hose behind turntable. *JFS/WMC*

In 1968, *Snorkel Fire Equipment Co.* introduced a new type of aerial master stream device. The *Squrt* was a 50-foot, two-section articulating boom with a remote-controlled master stream nozzle instead of a basket at its tip. Following a national demonstration tour, prototype *Squrt* on Ford C-Series chassis was sold to Indianapolis, Indiana. *WMC*

The Detroit Fire Department retrofitted two 1967 Mack C-Series pumpers with *Squrt* booms in 1969. Taking up relatively little space in the hose bed, a *Squrt* master stream device greatly enhanced any pumper's versatility. *WMC*

Maxim Motor Company introduced two all-new custom fire apparatus chassis designs in the early 1960s. The company's first *cab-forward* model was the clean-lined, flat-faced *Model "F."* Bristol, Connecticut, got this Maxim Model "F" tractor-drawn aerial with 100-foot ladder, booster equipment and open semi-cab in 1968. *WMC*

Maxim's other custom was the snub-nosed *Model "S"* conventional, which replaced the classic long-nosed Maxim custom built from 1946 through 1959. Like the companion cab-forward Maxim *"F,"* the new "S" was available in both tractor-drawn and four-wheel service aerial configurations. This sturdy-looking 100-foot aerial was delivered to Louisville, Kentucky, in 1963. *WMC*

Despite introduction of its own custom cab-forward fire apparatus chassis in 1961, Pirsch continued to offer its seemingly timeless long-nosed conventional pumpers and aerials well into the 1970s. Lexington, Kentucky, ordered this Pirsch *Type 41 Senior* 100-foot, tractor-drawn aerial in 1968. By now Pirsch customers could also specify a four-section, 100-foot ladder with fixed tiller seat. *MLC*

Pirsch customers could save some money on aerial truck purchases by specifying commercial-truck tractor instead of premium-model Pirsch Custom. Atlanta, Georgia, and Cleveland, Ohio, purchased a number of *Pirsch Senior* 100-foot aerials with Ford C-Series tractors and extended canopy-style cabs. This one went to Atlanta in 1967. *Roger Bjorge*

Smeal Mfg. Company of Snyder, Nebraska, started making aerial apparatus in the mid-1960s. Typical of early Smeal ladders was this 65-foot aerial on tilt-cab 1969 International Cargostar chassis for Chadron, Nebraska. Note the handy seat for the aerial ladder operator overhanging the rear of this compact apparatus. *WMC*

American-LaFrance delivered four *900 Series* 100-foot tractor-drawn aerial ladder trucks to Detroit, Michigan, in 1969. Note the four hydraulic outriggers — two on the turntable and two more below the booster hose reel. Two years after its infamous riots, Detroit was still ordering aerials with wide-open tiller seats. *WMC*

American LaFrance delivered some impressive tandem-axle aerial platform quints in the early 1970s. A fine example is this 900 Series *Aero-Chief 90* with 1000 gpm pump for North Olmstead, Ohio. Note dual outriggers behind rear wheels. *JFS/WMC*

CHAPTER EIGHT: THE 1970s

Just as the elevating platform revolutionized aerial firefighting in America in the late 1950s and through the 1960s, the *rearmount aerial* swept the industry in the 1970s. As the decade began, every one of the industry's principal players – *American LaFrance, Seagrave, Pirsch, Maxim, Mack, Ward LaFrance* and *Crown* – offered compact rearmount aerials on their proprietary custom chassis. Nearly all of these had single rear axles and 100-foot, four-section aerial ladders.

The telescopic-boom aerial towers *Sutphen* and *Mack* introduced in the mid-1960s were beginning to seriously challenge the articulating-boom type which had sparked the elevating platform revolution a decade earlier. In 1970, Mack began using Baker booms on its 75-foot *Aerialscope,* on the new wide-cab Mack *CF-600* custom fire apparatus chassis introduced in 1968. Over the next two decades, New York City (FDNY) purchased hundreds of these rugged Mack/Baker *Tower Ladders.*

In the early 1970s, American LaFrance's parent *A-T-O Corp.* acquired the Snorkel Fire Equipment Co. of Grandview, Missouri – makers of the Pitman-type Snorkel that had started it all, and the articulated *Squrt* water tower boom. By 1970, Snorkel had replaced its original 50-foot Snorkel with a much-improved 55-foot version.

At about this same time, Snorkel introduced an entirely new product which would quickly become the division's most prolific. The *Tele-Squrt* was a combined aerial ladder/water tower with a two-section 50-foot ladder mounted atop the telescopic master stream boom with integral piped waterway and remote-control master stream nozzle. A pair of A-frame outriggers behind the rear axle stabilized the unit when the boom was raised. The rearmount *Tele-Squrt's* pedestal took up very little space in a pumper's hose bed. What's more, the *Tele-Squrt* could be easily retrofitted to any older pumper. The new *Tele-Squrt* vastly enhanced any pumper's usefulness and made an aerial ladder/water tower readily available to even the smallest fire departments.

Independent aerial ladder maker *Grove* was beginning to come on strong. In 1970, Chicago purchased five Ward LaFrance rearmount aerials equipped with Grove 100-foot four-section ladders, and in 1972 the

In 1968, Mack replaced its aging *C-Series* with the all-new *"CF"* custom cab-forward fire apparatus chassis. From 1970 on, all *Mack Aerialscopes* were built on wide-cab *CF-600* chassis with 75- or 95-foot *Aerialscope* booms made by Baker Engineering & Equipment Co. of Richmond, Virginia. Over the next two decades, FDNY bought more than one hundred of these rugged Mack/Baker Tower Ladders. *WMC*

Boston Fire Department ordered two of these special *Squrt* rigs which ran as hose wagons for two-piece engine companies. Maxim built the bodies, which were mounted on 1971 Ford C-Series tilt-cab chassis. Here's Engine 39's *Squrt. JFS/WMC*

company introduced its first aerial tower. Grove fitted a crew basket and integral waterway to its rugged three-section 85-foot aerial ladder – the basket overhanging the cab on the tandem-axle, rearmount apparatus. Mounted on an Oshkosh custom cab-forward chassis, the first Grove Aerial Tower was delivered to Pine Castle, Florida.

There were a number of other notable product developments in 1972. Seagrave announced its *Astro-Tower* – an articulating-type elevating platform on the company's low-profile "S" series custom chassis. Equipped with a 70-foot *Trump Snorkel,* just one of these single rear axle towers was actually ever delivered – to Greenville, South Carolina. At the opposite end of the spectrum, *Pierce Mfg.* introduced its *Aerial-Ett* – a 35-foot utility type junior aerial platform. Only about a dozen were sold, all on light commercial truck chassis. And *American Fire Apparatus* came out with its *Aqua-Jet* – a telescopic master stream device with a ladder on top, a la Snorkel's *TeleSqurt.*

Also this year, Hendrickson Mobile Equipment introduced its 1871 Series cab-forward fire apparatus chassis, which subsequently became quite popular with makers of aerial devices. Initially sold through International Harvester, the 1871-S sported a distinctive, multifaceted slab-sided cab. Significantly, the model number reflected the year of the great Chicago fire.

Held in Cleveland, Ohio, the 1972 International Association of Fire Chiefs Convention featured an impressive array of aerial equipment. *Ward LaFrance* introduced its *Tele-Boom* – a midmount aerial platform with Reading Techmatic 75-foot telescopic boom, and the *Command Tower* – a nozzle-quipped platform that rose vertically 22 feet above the WLF custom pumper's midship-mounted pump. *American Fire Apparatus* exhibited a big quint on Oshkosh chassis with Grove 85-foot Aerial Tower. *American LaFrance* and *Mack* both exhibited custom-chassis pumpers equipped with 55-foot Snorkels.

With its purchase of *Snorkel* the previous year, American-LaFrance now offered an entire family of Snorkel-equipped products on its 900/1000 Series custom chassis; entry-level *Pioneer* and new mid-range *Pacemaker* chassis. Pumpers with 55-foot Snorkels were called *Super Chief Juniors;* Squrt-equipped pumpers *Aqua-Chiefs* and those with Tele-Squrts, *Duo-Chiefs.* In 1973, ALF discontinued production of its proprietary *Aero-Chief* articulating platform, replacing it with 65-, 75- and 85-foot *Snorkel* booms.

Also in 1972, the Calavar Corporation delivered its first 150-foot *Firebird* elevating platform to Tucson, Arizona. *Calavar Firebirds* were now available in three sizes – 90, 125 and 150 foot on special Seagrave or Hendrickson chassis.

Maxim Model "F" aerials were available in straight-frame service or articulated tractor-trailer types with closed or open semi-cabs. Quincy, Massachusetts' Ladder 3 responded with this 1971 Maxim 100-foot service aerial. Note traffic-busting air horns above windshield. *JFS Photo courtesy of Gregg D. Merksamer*

Pitman Snorkels were mounted on virtually all custom fire apparatus chassis, including *Maxim, Pirsch, FWD, Mack, Crown* and *Seagrave.* Peter Pirsch & Sons built this 85-foot Snorkel Quint on its own cab-forward custom chassis — including the body. Most other Snorkels of this era had Pierce-built bodies. Franklin Park, Illinois' 1971 Pirsch Snorkel has Chicago-style black upper cab. *MLC*

A Canadian fire apparatus manufacturer – *Pierre Thibault Ltd.* of Pierreville, Quebec – had entered the lucrative U.S. aerial truck market in the mid-1960s. In 1972-73, Thibault delivered six tractor-trailer 100-foot aerials to the Los Angeles City Fire Department. Drawn by Duplex-chassis tractors (three open and three closed) these big tillered aerials had four-section 100-foot ladders and fixed tiller seats. But there were continuing problems with the Thibault ladders – all of which were later replaced with Seagrave 100-foot aerials. On various custom and commercial chassis, Thibault aerial ladder trucks were purchased by a number of other American fire departments including St. Paul, Minnesota, San Diego, California, and Columbus, Ohio. Hahn and Ward LaFrance also built aerials equipped with Thibault 85-foot and 100-foot ladders.

In 1973, Grove Mfg. sold its aerial ladder division to a group of investors led by Mahlon Zimmerman – who formed a new company called *Ladder Towers Incorporated,* or *LTI.* The following year LTI began making the former Grove aerial ladders and aerial towers in a new facility in the Conestoga Industrial Park in Leola, Pennsylvania. Before long, this aggressive new company was building complete aerial apparatus in a separate plant in Ephrata, Pennsylvania. In 1974 LTI introduced its first all-new product. The *LTI Aqua-Stix* was a lightweight 75-foot, three-section aerial with piped waterway and a remote-control ladder pipe at its tip.

At the 100th Annual Convention of the International Association of Fire Chiefs (IAFC) held in Baltimore, Maryland, in the fall of 1973, American LaFrance introduced the successor to its ageing 900/1000 custom fire apparatus chassis. The new *Century Series* custom featured a full-width cab with huge curved windshield. The new *Century Series* product line included midship and rearmount service aerials and quints, tractor-drawn, tillered aerials, and *Snorkel* elevating platforms.

In 1974, the Chicago Fire Department commissioned a unique one-off aerial truck equipped with a rear-mounted 136-foot *MoritaLift* aerial ladder made in Japan. Mounted on a special Hendrickson tandem-axle chassis with a three-man cab and Pierce bodywork, the six-section *Morita* ladder was equipped with a cable-operated rescue basket designed to whisk rescued citizens or firefighters safely and quickly to the ground. Assigned to Chicago's busy Hook & Ladder Co. 2 in the downtown loop area, the *MoritaLift* aerial had a relatively short service life. No more were purchased.

Newcomer *LTI* quickly made its mark on the independent aerial ladder business. In 1975 the company introduced a compact 65-foot aerial ladder utilized by the Young Fire Equipment Corp. of Buffalo, New

Seagrave Corp. had two arms-length "cousins" — *Clintonville Fire Apparatus* located in corporate parent FWD's home city of Clintonville, Wisconsin, and *Seagrave-Timpco* (*Timmons Metal Products*) in Columbus, Ohio. Fort Smith, Arkansas, got this 1972 *CFA/Seagrave* on canopy cab Ford C-Series chassis with Seagrave 100-foot rearmount aerial ladder. *RMA/WMC*

One of the most photogenic rearmount aerials around had to be College Park, Maryland's 1972 Pirsch cab-forward custom with two-tone paint and cut-down windshield, which permitted lower travel height for the four-section, 100-foot ladder. Note the bell and siren on the front bumper extension. *Alan M. Craig Photo*

York, on its compact short-wheelbase *Fire Commander* quint. Also in 1975, innovative LTI announced the *Streamliner* – a futuristic-looking rearmount aerial or platform tower on a special ultra-low profile Oshkosh tandem-axle chassis, and the *Astro Tower* (not to be confused with the Seagrave tower of the same name) – a jaw-dropping highrise rescue unit with a maximum vertical reach of 240 feet!

Exhibited at *Fire Expo 75* in Lancaster, Pennsylvania, the *LTI Astro Tower* consisted of a four-section, telescopic base tower – much like a construction crane – with an 85-foot or 100-foot LTI aerial ladder mounted on a cantilevered platform atop the vertical base section. But the daring *Astro Tower's* 20-minute setup time and equally stratospheric price blunted its appeal. As far as is known, none were ever sold for fire service.

Less than a decade after its introduction in the mid-1960s, the *Sutphen Aerial Tower* had been enthusiastically received by the American fire service. Early Sutphen towers – mostly 85-footers – were mounted on heavy commercial truck chassis – Fords, GMC and the semi-custom International CO-8190. Tandem rear axles soon became standard. Sutphen introduced its own custom chassis in 1972. Sutphen Towers were in front-line service on fire departments large and small all over the country. Big-city users included Boston, Washington, D.C., Detroit and Cleveland. Syracuse, New York, and Knoxville, Tennessee, replaced most of their conventional aerial ladder trucks with Sutphen Towers. Most of the 85 and 100+ towers being delivered by now were also equipped with 1000 or 1250 gpm pumps.

Until the introduction of fixed, prepiped waterways in the early 1970s, it had been standard fire department practice to run three-inch hose up the aerial ladder when the ladder was to be used as a water tower. The ladder hose fed a fixed or demountable ladder pipe which was aimed by a firefighter standing at the tip of the ladder. Not long after its new *Century Series* apparatus went into production, American LaFrance in 1976 introduced a 100-foot rearmount aerial with prepiped waterway called the *Water Chief.* Telescopic pipes on either side of the ladder fed a power-operated, remote-control monitor that was permanently attached to the top section of the ladder. The new *Water Chief* was an effective heavy-duty combination aerial ladder and water tower.

American LaFrance in 1979 introduced a smaller, lighter version of the *Water Chief* with a three-section 75-foot ladder. Mounted on a standard *Century Series* custom-chassis pumper, the *Water Chief 75* quickly became one of the company's most popular products. Like its larger, heavier namesake the *Water Chief 75* had a side-mounted, prepiped waterway and fixed, remote-control master stream nozzle at its tip.

By the early 1970s, Peter Pirsch & Sons Co. had joined American LaFrance and Seagrave in offering tractor-drawn, four-section 100-foot aerial ladder trucks with fixed tiller seat. Boston suburb of Brookline, Massachusetts, got this imposing 1972 *Pirsch Senior* 100-foot custom aerial with enclosed tiller cab. *MLC*

American LaFrance introduced its *Ladder Chief* rearmount in 1968. *Ladder Chief* was also available as a quint with high-capacity pump. West Seneca, New York, near Buffalo got this premium-model 1972 ALF *1000 Series Ladder Chief* quint with notched rear cab roof to minimize travel height of the100-foot rearmount aerial. *Alan M. Craig Photo*

In the late 1970s, *Snorkel* again expanded its already diverse family of aerial devices with a new "junior" member. The *Squrt 35* was a miniaturized version of Snorkel's 50-foot articulating-boom *Squrt.* With a working height of 35 feet, this new junior master stream device was designed for installation on a standard pumper, new or old.

But the most significant development of the 1970s was the stunning rise of two relative newcomers to the industry. In the late 1960s and through the '70s, *Pierce Mfg.* of Appleton, Wisconsin, grew from a small regional builder into one of the industry's major national players. Formed in 1974, *Emergency One* of Ocala, Florida, came out of seeming nowhere to the front rank of the industry by throwing away the rule book and building fire trucks its own way, utilizing modular aluminum bodies. But it wasn't until the 1980s that these two upstarts became serious challengers in the sophisticated aerial devices market.

In 1979, fast-growing *Emergency One* – which had moved into a spacious new plant in Ocala, Florida, the previous year – was purchased by the *Federal Signal Corp.* of Chicago. That same eventful year *Emergency One* entered the U.S. aerial truck market with the 106-foot *Fire Spire* aerial ladder originally developed by National Foam and supplied on an OEM basis to several independent fire apparatus builders by the Hahn Motor Co. of Hamburg, Pennsylvania.

As the 1970s ended, one was likely to find a variety of aerial devices in use at major fires. While the elevating platform had fundamentally changed aerial firefighting in North America, it had *not* replaced the tried-and-true aerial ladder. Articulating elevating platforms and telescopic aerial towers typically worked alongside conventional aerials. *Squrts* and *TeleSqurts* also poured water onto the fire. Once they got used to them, most firefighters preferred the smooth, effortless ride of an elevating platform basket to clambering up and down an aerial ladder.

Calavar Corp. took the industry by surprise with its 125-foot *Calavar Firebird* aerial platform in 1969. Three years later, Calavar introduced a *150-foot* platform — highest in the U.S. First *Firebird 150* went to Tucson, Arizona, in 1972. Note huge front tires on this monster tower with hydraulic drop-down side ladder beds. *JFS/WMC*

U.S. fire apparatus manufacturers showcased their latest wares at the 1972 IAFC Convention in Cleveland, Ohio. Among rigs exhibited on streets around the convention hall was this unusual *American Fire Apparatus Co.* quint with pump control panel on front of cab and Grove 100-foot aerial ladder ready for delivery to North Aurora, Illinois. *JFS/WMC*

Ward LaFrance championed the switch from red to lime-green and safety-yellow fire apparatus in the early 1970s. Studies showed lime green/yellow was much more visible at night. Batavia, New York, got this white over lime yellow 1973 Ward LaFrance *Ambassador* rearmount quint with *Ladder Towers Incorporated* (LTI) 100-foot ladder. *Alan M. Craig Photo*

By the early 1970s, *Sutphen Corporation* had developed its own tandem-axle custom chassis with Duplex canopy-style cab. Here's a profile shot of an 85-foot *Sutphen Aerial Tower* with pump then in service in Wayne Township, New Jersey. Note the "huck-bolted" four-section telescopic box boom and low-hanging crew bucket with side-mounted monitor nozzle. *JFS/WMC*

Mack found many takers — besides New York City — for its rugged *Aerialscope* Tower Ladder. Owned by Greenville, New York, this well-dressed 1973 *Mack Aerialscope* with Baker 75-foot boom on CF-600 chassis was photographed at big SPAAMFAA antique fire apparatus muster held at Kenisco Dam Plaza in Valhalla, New York, in June 1976. *Richard J. Allen*

The civil strife that engulfed America's largest cities in the late 1960s accelerated the trend toward fully-enclosed fire apparatus. In 1972, the Cincinnati Fire Department ordered several "riot-proof" Seagrave pumpers and aerials with four-door sedan cabs and enclosed tiller seats. Note the absence of a window between the doors on Truck 12's rig. *Steve Hagy Photo*

Ward LaFrance Truck Corp. announced its *Command Tower* elevating platform at the 1972 IAFC convention. The *Command Tower* platform with integral deck turret rose vertically 22 feet above pump panel. Note the telescopic ladder on the right-hand side of Buffalo, New York's Engine 13, and triple-jointed piping for the deck monitor. *WMC*

Grove introduced its first aerial tower — essentially the company's sturdy steel aerial ladder with a crew basket and pre-piped monitor at its tip — in the early 1970s. The American Fire Apparatus Co. of Battle Creek, Michigan, built this big rearmount tower with Grove 85-foot platform for Fairfax, Virginia, in 1974. *Shaun P. Ryan Photo*

The *Chicago Fire Department* placed this one-of-a-kind "high ladder" into service in 1974. Built on Hendrickson tandem-axle chassis with special three-man cab and Pierce bodywork, it was equipped with a Japanese-made, rear-mounted 136-foot *MoritaLift* aerial ladder with two-man rescue basket that ran up the center of the ladder. Ground ladders were stacked on the open deck. A noble experiment, it was, however, something less than a success and no more were ordered. *Dan G. Martin Photo*

Canadian fire apparatus manufacturer *Pierre Thibault* was aggressively marketing its own aerial ladders in the United States. Big-city customers included San Diego and Los Angeles, California, and St. Paul, Minnesota. Columbus, Ohio, ordered four of these tillered Thibault/Mack CF 100-foot tractor-drawn aerials in 1974. *WMC*

Squrt articulated master stream devices were often mounted on fire support vehicles — hose wagons, auxiliary power and lighting apparatus, etc. Grand Rapids, Michigan, ran *Snorkel Generator 1* — a 1974 Pierce/Ford power and lighting unit equipped with a 54-foot *Squrt* boom. *Dan Jasina*

Following announcement of the all-new *Century Series*, American LaFrance delivered its last 900/1000 Series fire apparatus in 1974 and early 1975. Bedford Heights, Ohio, got this 1975 *1000 Series Ladder-Chief* 100-foot rearmount aerial. Note single rear axle, high angle of four-section ladder and easily accessible stacked ground ladders. *JFS/WMC*

More and more aerial ladder trucks and elevating platforms were being ordered with full-sized fire pumps. Stow, Ohio, ordered this Pirsch Custom Cab-Forward 85-foot service aerial with 1000 gpm pump and tandem rear axles in 1975. Note the preconnected soft suction hose on the front bumper extension. *JFS/ WMC*

Once a small builder of rural and small-town pumpers, the *Boardman Company* of Oklahoma City, Oklahoma, was building ever larger, more sophisticated apparatus. Bristow, Oklahoma, ordered this 1975 Boardman/ Duplex Quint with 1000 gpm pump, 300-gallon tank and *LTI* 85-foot tower. *Steven B. Loftin*

California's state capital of Sacramento had more *Calavar Firebirds* — four in all — than any other city in North America. Truck 2 was assigned this 1975 *Firebird 150* on Hendrickson chassis, one of two on the S.F.D. roster. *Chuck Madderom*

Seagrave hit a real home run with its industry-leading *Rear Admiral* rearmount aerial in the mid-1960s. By the 1970s, every major fire apparatus manufacturer had introduced its own version of the compact, space-efficient rearmount. Indianapolis, Indiana, assigned this 1976 Seagrave *Rear Admiral* with "P" cab and four-section, 100-foot ladder to Aerial Co. 27. *Rodger Birchfield*

The *Boston Fire Department* purchased ten of these Seagrave tractor-drawn 100-foot aerials with four-door sedan cabs and fully enclosed tiller seats in 1976. Tiller cabs were equipped with sliding doors and plastic "blister" windows. Note road sanders under the tractor. *BFD Photo/Bill Noonan*

The Maxim Motor Company, which had advanced the rearmount aerial concept with its short-wheelbase *Maxim/Magirus* rearmounts two decades earlier, introduced a next-generation version on its *Model "F"* cab-forward custom chassis. Cranston, Rhode Island, got this chrome-yellow 1976 Maxim Rearmount on tandem-axle "F" chassis with 100-foot, four-section Maxim ladder. *L. Murray Young Photo*

All three types of aerial devices in general use in the 1970s are seen here in action at a five-alarm apartment building fire in downtown Detroit. At left is Ladder 6's 1970 *Sutphen Aerial Tower*. Ladder 1's 1965 *FWD/Pitman Snorkel* directs stream into sixth-floor window while three conventional aerial ladders with ladder pipes are in use in foreground. *WMC*

Like a number of other cities, Jackson, Michigan, retrofitted one of its front-line pumpers, a 1977 American-LaFrance *Century Series,* with a *Squrt 35* master stream device. *WMC*

How *low* can they *go?* Not much more than this! Oshkosh developed an ultra-low-profile chassis especially for mounting aerial devices. *Young Fire Equipment* of Buffalo, New York, delivered this low-slung 1976 *Streamliner* Quint with *Ladder Towers Incorporated* (LTI) 100-foot rearmount aerial to Snyder, New York. *WMC*

Detroit switched from red to trendy new lime-green in 1973. Ladder 10 was assigned this 1977 Seagrave 100-foot tractor-drawn aerial with water tank and dual booster hose reels on the tractor, and enclosed tiller cab. Upper cab and four-section aerial ladder were painted white. *WMC*

New York City ordered a dozen of these long Seagrave tractor-drawn 100-foot aerials with four-door safety cabs in 1974. Note also the utilitarian painted front bumper and plain siren mounted on the face of the cab and sliding tiller seat enclosure. *MLC*

Once a novelty, fully-enclosed tiller cabs and four-door tractors became requisites for crew safety on aerial ladder trucks in big cities like New York and Boston — both of which ordered Seagrave tractor-drawn aerials like the one seen here, one of 12 delivered to the FDNY in 1974. Note the sliding rear cab enclosure and short attic ladder. *MLC*

The FDNY also ordered a substantial number of Seagrave 100-foot rearmount aerials with four-door sedan cabs. Note the stand-up "telephone booth" crew position amidships, and the protective screen over the cab side window — a sad commentary on the times. New York truckies still used pompier (scaling) ladders — two of which are mounted above the ground ladders. *WMC*

Reading Techmatic of Reading, Pennsylvania, introduced the *Readi-Tower* — a four-section telescopic aerial tower device with emergency escape ladder mounted atop the 75-foot boom. Ward LaFrance delivered this *Tele-Boom* with 1000 gpm pump on Ford C-Series chassis to Louisville, Ohio, in 1977. *JFS/WMC*

Peter Pirsch & Sons started making this distinctive cab with notched rear in the early 1960s. By the late 1970s, this costly, labor-intensive design was nearing the end of its production cycle. Louisville, Kentucky, received this 1977 Pirsch midmount service aerial with four-section 100-foot ladder. Note the radio speaker atop the control pedestal. *MLC*

During the 1970s, Pirsch delivered some aerials on Mack CF-600 custom chassis. They included midmount, tractor-drawn and rearmount aerials like this one for North Riverside, with 100-foot aerial and pump. *MLC*

American-LaFrance redesigned its steel aerial ladder in the late 1970s. Visible change included redesigned ladder lift cylinders, which angled *forward* instead of to rear as on earlier models. Most aerial trucks now had flop-down hydraulic stabilizers like the ones visible under the turntable. This *Century Series* 100-foot service aerial with booster equipment and fixed ladder pipe went to ALF's home city of Elmira, New York. *WMC*

One of New York City's famed *Mack Aerialscope* Tower Ladders swings into action at a spectacular warehouse blaze in Brooklyn, New York, on June 17, 1976. Powerful ladder pipe at right provides cooling effect as flames boil from second- and third-floor windows. Note yet-uncharged hose lines hooked into Ladder 119. *Alex Donchin Photo*

Snorkel Fire Equipment introduced what was destined to become its most successful and prolific product in the early 1970s. The *TeleSqurt* was a two-section, telescopic water tower with a low-handrail, light-duty aerial ladder mounted on top. Thousands of 55-, 65- and 75-foot *TeleSqurts* have since been delivered and continue to be sold to this day. Pierce built this 1000 gpm pumper on tandem-axle 1977 Ford C-Series chassis with 55-foot *TeleSqurt* for Guildford County, North Carolina. *JFS/WMC*

American LaFrance introduced its *Water Chief* — a combination aerial ladder/water tower with pre-piped waterway — in 1976. Cleveland, Ohio, got this ALF *Century Series* 100-foot *Water Chief* originally ordered by Mexico City in 1979. Telescopic waterway was mounted on sides of the four-section ladder. Note also the well-protected remote-control monitor at tip. *JFS/WMC*

Crown Firecoach of Los Angeles introduced a new, full-width cab in 1977. Arcadia, California, got the only elevating platform built on the second-generation *Crown Firecoach* chassis. Truck 31 sports an 85-foot *Snorkel* boom and Pierce bodywork. *Chuck Madderom*

Ladder pipes atop four Seagrave aerial ladders and a Pitman Snorkel make short work of an extra-alarm fire in Detroit. Unusual close-up of aerial action was made from adjoining building. *DFD Photo*

Four compact rearmount aerials — three Macks and a Thibault — in action as water towers at a major office building fire in downtown Montreal, Quebec. With many narrow old downtown streets, Montreal ran two-piece ladder companies consisting of a short-wheelbase rearmount aerial accompanied by a tractor-drawn city service ladder truck. *WMC*

Long Island, New York, fire companies were renowned for their eye-popping, cost-is-no-object fire apparatus. Here's a good example. Manhasset-Lakeville ordered this humongous Grumman-built, tractor-drawn aerial with towering Kenworth canopy cab tractor and *LTI* 100-foot aerial. *Rod Hodgson*

Walpole, Massachusetts, ordered this 1980 Pirsch rearmount aerial on custom tandem-axle chassis with TCM "Cincinnati" cab instead of Pirsch's proprietary cab-forward design. Aerial is Pirsch's distinctive four-section 100 footer. Note single set of outriggers at extreme rear of apparatus — which also has 1250 gpm pump and 300-gallon tank. *MLC*

CHAPTER NINE: THE 1980s

Competition in the aerial devices market intensified during the 1980s with the inevitable entry of two bold, new players – *Emergency One* and *Pierce.* Once the province of only the industry's largest members – American LaFrance, Seagrave, Pirsch, Maxim, Mack, etc. – aerial ladder trucks and platforms were now also made by some of the smaller, second-tier producers like Grumman, Hahn, Bean/FMC, American – *and* independent aerial ladder maker LTI.

The *quint* concept had firmly taken hold, and many aerial ladder trucks were now being ordered with midship-mounted full-size (1000-1250 gpm) pumps, necessitating heavier chassis and tandem rear axles. The 1980s also saw a marked shift away from open rear-entrance canopy-style cabs to four-door sedan cabs – a trend sparked by civil unrest in the nation's major cities and increasing concern for the safety of firefighters responding to and returning from alarms.

Founded just six years earlier, *Emergency One* introduced a host of increasingly ambitious aerial devices through the 1980s. Following its initial foray into the aerial market with the Hahn *Fire Spire* ladder in 1979, the Ocala, Florida, company fielded its first proprietary aerial device in 1980. The *Strato-Spear* was a 50-foot combination water tower and aerial ladder similar in design to Snorkel's extremely successful *TeleSqurt.* Soon after, *Emergency One* also began offering its customers the *Simon Snorkel* – a 103-foot, three-boom elevating platform with unique side-mounted escape ladder made by Simon Engineering in England. Most of these were mounted on tandem-axle Hendrickson 1871 chassis.

In 1981, Sutphen delivered two 100+ *Sutphen Aerial Towers* to the New York City (FDNY) Fire Department – which up to now had purchased Mack/ Baker *Aerialscope* tower ladders almost exclusively. Three years later, the FDNY purchased two massive

Introduced in 1979, American LaFrance's *75-foot Water Chief* became one of the company's most popular products. Essentially a triple combination pumper equipped with a three-section, 75-foot aerial ladder with piped waterway, the *Water Chief 75* found great favor with smaller fire departments. Northville, Michigan, got this 1981 ALF *Century Series Water Chief 75.*

Cleveland, Ohio, modernized with a big fleet of new American-LaFrance pumpers, aerials and snorkels in 1980. Among them was this *Century Series* 100-foot *Ladder Chief* quint with 1500-gallon-per-minute pump. Note the two fire extinguishers carried on front bumper. *JFS/WMC*

low-profile American LaFrance tower ladder quints equipped with 100-foot *LTI* aerial towers and 1500-gallon-per-minute pumps.

Mack, meanwhile, had introduced its *Bulldog 1* – a compact rearmount aerial with 106-foot LTI aerial ladder on Mack's custom CF-611 pumper chassis. Mack Trucks had also supplemented its standard 75-foot *Aerialscope* with a heavier 95-foot version.

In 1982, *Emergency One* entered the major leagues of aerialdom with an extruded aluminum aerial ladder of its very own design and manufacture. The rear-mounted, four-section 110-foot *Strato-Spear* was a real success, followed one year later by the company's first proprietary aerial platform, the 95-foot *Strato-Spear,* also of welded extruded aluminum construction. In 1984 – the same year the company introduced its *Hurricane* custom chassis – *Emergency One* also came out with a five-section *Strato-Spear* 135-foot rearmount aerial – the tallest ladder made in America at the time.

One of the industry's other relatively new players was *Grumman Allied Industries* of Bethpage, Long Island, New York, which had entered the fire apparatus industry in 1976 with its acquisition of *Howe Fire Apparatus* of Anderson, Indiana. Howe had earlier purchased *Oren Roanoke* of Roanoke, Virginia, and *Coast Apparatus* of Martinez, California. At first, *Grumman Emergency Products* marketed aerial apparatus with LTI, Hahn, Snorkel and Readi-Tower aerial devices, but in 1982 introduced its own *AerialCat* – a 95-foot aerial tower developed in conjunction with *Steeldraulics.* A 102-foot *AerialCat* platform followed three years later.

Late in the decade Grumman introduced a 121-foot steel *AerialCat* rearmount aerial which boasted the industry's tallest piped-waterway. Most of Grumman's *AerialCats* were built on Duplex chassis – some with distinctive raised rear cab roofs notched to accommodate the ladder/tower boom in its travel position. Introduced in 1988, Grumman's new *Panther* aerial platform was mounted on a special low-profile chassis supplied by *HME.*

The industry's other powerful new player – *Pierce Mfg.* – had offered *LTI* aerial ladders and platforms on its fire apparatus since the mid-1970s. In 1984, Pierce entered into an arrangement to source its aerial ladders from the *Smeal Mfg. Co.,* of Snyder, Nebraska. This mutually beneficial arrangement was in place for eight years, until Pierce began making its own aerial ladders and platforms in 1992.

After more than four decades, *Peter Pirsch & Sons Co.* extensively redesigned its trademark aluminum alloy, lattice type "senior" aerial ladder. Introduced in 1983, Pirsch's new *Skytop* 110-foot rearmount aerial featured noticeably thicker, heavier structural members. Rival *Seagrave* also introduced a new heavy-duty aerial ladder design – the *RA110* introduced in 1987.

After 153 years, longtime industry leader *American LaFrance* closed its plant in Elmira, New York, in mid-1985. But corporate parent company *Figgie International* set up a new American LaFrance operation in Bluefield, Virginia, the following year. At first the born-again company offered only pumpers on its new *Century 2000* custom chassis, but two years later the company was again building aerial ladder trucks – midship-mount service aerials and quints, big tractor-drawn tillered aerials – and a second-generation *Water Chief II* 75-foot aerial/pumper. Designed in the late 1930s, ALF's three- or four-section steel aerial got a redesigned base section with heavier butt-end girts and reinforcing.

Since the arrival of Seagrave's industry-leading *Rear Admiral* in the mid-1960s, the compact rearmount had emerged as the most prolific aerial ladder truck configuration in North America through the 1970s and `80s. While most manufacturers continued to also offer conventional midship-type service aerials and quints, the venerable tractor-drawn, tillered aerial had steadily fallen out of favor. New York, Los Angeles and San Francisco were the only major cities still routinely ordering tractor-drawn aerials in any numbers. With its huge fleet of Mack tower ladders and steadily-increasing numbers of Seagrave and American LaFrance rearmounts, the traditional tractor-drawn aerial hook and ladder was less and less a common sight on Gotham's streets. But in some congested neighborhoods with ancient firehouses and crowded, narrow streets, nothing else would do!

As the 1980s drew to a close, the U.S. fire apparatus industry had entered a period of turmoil and restructuring not seen since the automobile replaced the horse. Intense competition from a host of aggressive new companies unencumbered by decades of tradition – *E-One, Pierce* and *LTI* – was taking a mounting toll on the grand old names of the industry. One by one, some of the industry's oldest and best-known players were falling by the wayside. *Maxim* ceased production in late 1989, as did *Hahn. Pirsch* went down for the count for the second time in three years. The last *Crown Firecoach* was built in 1985, and *Mack* – which had delivered its last complete factory-built fire apparatus in 1984 – discontinued production of its CF custom fire apparatus chassis, exiting the industry for good. Once-dominant *American LaFrance* was little more than a shadow of its former self.

Snorkel Fire Equipment's 54-foot *Squrt* boom was ideally suited for mounting on hose wagons. Lexington, Kentucky, ran this Squrt-equipped hose wagon on flat-faced 1980 Mack MB chassis. Note huge windshield. *Steve Hagy Photo*

Fourth in the Snorkel family of aerial master stream devices was the *Squrt 35* — a midship-mounted junior version of the company's articulated 54-foot *Squrt*. Boardman Company of Oklahoma City, Oklahoma, built this pumper on tilt-cab International Cargostar chassis with 35-foot *Squrt* for Little Rock, Arkansas. *JFS/WMC*

One of the smaller regional builders, *Sanford Fire Apparatus* of East Syracuse, New York, built some *big* fire apparatus — including this aerial tower on 1980 Oshkosh A-1838 custom chassis with *LTI* 85-foot platform for Montrose, New York. *WMC*

Mack Trucks edged back into the crowded U.S. aerial ladder market again in the early 1980s. In mid-1981, Mack announced the *Bulldog 1* — a 106-foot rearmount aerial on Mack CF custom chassis with Hahn *Fire Spire* ladder. Mack also advertised midship-mount and tractor-trailer types. *MLC*

By the early 1980s, Pirsch had stopped making its own cab-forward cabs and began using the familiar standard cab made by *Truck Cab Manufacturers* (TCM) of Cincinnati, Ohio. The thinly disguised new Pirsch cab had distinctive protruding rectangular headlight bezels. This straight 100-foot rearmount aerial was delivered to Superior, Wisconsin.

Memphis Fire Department bought Pirsch aerials since the 1930s. Some of these well-used tractor-drawn aerials were sent back to the Kenosha, Wisconsin, plant for complete rebuilding — including new aerial ladder assemblies and tractors. In the early 1980s, Memphis sent this 1936 Pirsch aerial trailer back for total rehab including remounting a 1953 Pirsch 100-foot ladder and a new 1982 Pirsch/Ford L-Series tractor with canopy cab extension. *Richard M. Adelman Photo*

Founded only a decade earlier, *Emergency One, Inc.* of Ocala, Florida, emerged as one of the industry's leading players by the 1980s. In 1984, E-One introduced its first custom chassis — the *Hurricane* — and was making its own extruded-aluminum aerial ladders. The Boston Fire Department purchased six of these 1984 *E-One Hurricane* 110-foot rearmount aerials. *BFD/Bill Noonan Photo*

The Sutphen Corp. broadened its already comprehensive aerial product line with two new aerial devices — a 70-foot *Sutphen Mini Tower Quint* and a companion water tower. Boston's Engine 20 ran this 1984 Sutphen 1500-gallon pumper with 67-foot aerial ladder/water tower boom. *BFD/ Bill Noonan*

American LaFrance acquired *Snorkel Fire Equipment* in the early 1970s, immediately replacing its proprietary *Aero-Chief* elevating platform with *Snorkel* booms. ALF mounted all types of Snorkel aerial devices — including *Snorkels, Squrts* and *TeleSqurts* — on *Century* custom and commercial-chassis apparatus. Equipped with 55-foot Snorkel, this 1984 *Super Chief Junior* pumper was delivered to Ingomar Fire Company, McCandless, Pennsylvania. *WMC*

Industrial fire departments also purchased aerial platforms. Phillips Petroleum's works at Borger, Texas, were protected by this 1983 American LaFrance *Century Series* quint with 2000 gpm pump, 750-gallon foam tank and *Snorkel* 85-foot elevating platform. *Steven B. Loftin*

Sutphen Corporation complemented its famed 85- and 100-foot aerial towers with a junior edition called the *Sutphen Mini-Tower.* Port Huron, Michigan, operated this compact 1985 *Sutphen Mini-Tower* with 1500 gpm pump and 65-foot boom. Note the high handrails atop the boom. *Fred L. Beeton*

In 1982, American LaFrance moved into a new plant in the Elmira suburb of Southport, New York. Three years later, American LaFrance closed its doors. Among the rigs built in the new Southport plant was this *Century Series Ladder Chief* 100-foot rearmount aerial with low-profile canopy cab for Stony Brook, Long Island. *WMC*

Although well known around the world, relatively few English-built *Simon Snorkels* were sold in the United States and Canada. Strathcona County, Alberta, near Edmonton got this 103-foot *Simon Snorkel* built by Superior Emergency Equipment of Red Deer, Alberta, on 1982 International Cargostar tilt-cab chassis with 1250 gpm pump. *Dave Stewardson*

Seagrave introduced wide, new four-door cabs in the late 1970s. Cleveland, Ohio, got three of these big tandem-axle 100-foot rear-mount aerials with high side equipment compartments in 1985. The old style four-section ladder made since the 1930s would soon be history, replaced by the much more modern *Patriot.* JFS/WMC

Yet another new master stream device — the *Snozzle* — arrived on the scene in the 1980s. Developed by the *Crash Rescue Equipment Co.* of Dallas, Texas, *Snozzle* was originally intended for use on airport crash trucks. But more than a few were ordered for mounting on municipal fire apparatus, like this Snozzle-equipped pumper built by Boardman on 1986 International *S-Series* chassis. *Steve Loftin*

Steeldraulics, Inc. of Lancaster, Pennsylvania, built two of these 75-foot quints on 1985 Spartan Gladiator chassis for the San Francisco Fire Department — neither of which were ever placed into service with the SFFD, although two Steeldraulics/Spartan 100-foot tractor-drawn aerials were. *Chuck Madderom*

Ladder Towers Incorporated — better known as simply *LTI* — introduced a custom fire apparatus chassis of its own design. The *LTI Olympian* featured a large, sloping windshield and distinctive front wheel openings. Chili, New York, got this 1987 *LTI Olympian* with 100-foot aerial tower and 1750 gpm pump. *Gary Dinkel*

By 1987, born-again *Kersey/American LaFrance* was delivering pumpers and aerials on its redesigned *Century 2000* chassis. Wood-Ridge, New Jersey, got this *Water Chief Quint* with 2000 gpm pump and 100-foot aerial ladder. Note new structural reinforcement on tip of bed section. *WMC*

A factory-fresh Seagrave tractor-drawn 100-foot aerial photographed in the plant yard in Clintonville, Wisconsin. Note side-facing warning lights on cab roof, high side compartments and ladder for access to the flat-topped, enclosed tillerman's cab. *WMC*

Seagrave introduced this wide, new four-door cab design in the late 1970s. All four doors were of similar design, and the front-end featured practical *zip-out* panels for ease of maintenance. Here's a tandem-axle rearmount aerial with the now somewhat dated four-section 100-foot ladder. *WMC*

Seagrave announced a heavy-duty *RA110* aerial in the early 1990s. Note massive construction compared with Seagrave's standard aerial made since the 1930s. This RA110 with raised rear cab roof design sits in Clintonville, Wisconsin, plant yard all ready for its delivery trip. *MLC*

Kilgore, Texas, ran this 1988 Seagrave with heavy-duty 110-foot *RA* aerial ladder. Rear-entrance canopy-style cabs were relatively rare on new apparatus anymore, replaced by safer fully enclosed four-door safety cabs. *Steven B. Loftin*

Canadian manufacturer Thibault developed the *SkyArm,* an aerial platform with an articulated upper boom. When Thibault folded, Nova Quintech acquired the *SkyArm, SkyPod, SkyFive* and *SkyBoom.* Pierce bought them from NQ in 1997. Markham, Ontario, got this 1989 *Thibault/Pemfab SkyArm. WMC*

Successor to American LaFrance, *Kersey American LaFrance* of Bluefield, Virginia, introduced an all-new *Century 2000* custom chassis in 1987. Waterbury, Connecticut, got this 1989 American LaFrance *Century 2000* midship-mount 100-foot service aerial with four-door cab. *Neal A. Van Deusen*

Originally delivered to Forest Park near Cincinnati, Ohio, this like-new 1991 *Grumman AerialCat* with 1500 gpm pump and 102-foot tower/ladder was purchased by the Kingsville, Ontario, fire department in 2005. *WMC*

CHAPTER TEN: THE 1990s-2000s

For the motor fire apparatus industry, the 20th century ended much as it had begun – in a state of turmoil and restructuring. Many of the industry's oldest names were fast fading into history. *Maxim's* remaining assets were auctioned off in early 1990. *Mack* delivered its last *CF* custom fire apparatus chassis. *Ward 79* – corporate successor to *Ward* LaFrance – also threw in the towel this year. Beleaguered *Pirsch* – shut down twice in the 1980s – closed its doors for the last time in 1991. Reconstituted in rural Virginia in 1986, a vastly downsized *American LaFrance* was again struggling for survival. Inexplicably, in 1991 *Kersey/American LaFrance* acquired what was left of *Hahn*.

Much-stronger, heavier aerial ladders and a plethora of telescopic ladder-towers dominated the aerial industry as the 1990s began. The compact *rear-mount* remained the configuration of choice for aerials and quints – but the traditional *midship-mount* was making something of a comeback, as was the classic tractor-drawn tillered aerial. But the articulated elevating platform that had launched the aerial revolution in the 1950s and `60s was receding into the background. The last *Hi-Ranger* aerial tower had been delivered in 1989. Calavar Corp. had discontinued its mighty *Firebird* in 1979. Although *Snorkel Fire Equipment's Tele-Squrt* and *Squrt* aerial devices continued to sell well, few full-sized (65-, 75- and 85-foot) *Snorkels* were being sold anymore. Among the few articulating type platforms still available was the Finnish-made *Bronto Skylift* – North American distribution rights for which had been acquired by *E-One Inc.* in 1996.

Change rippled throughout the industry. *Baker*

Lime-green and white 1991 *LTI Olympian* 110-foot tractor-drawn aerial with low-profile four-door cab, Modesto, California. *Bill Hattersley*

The Detroit Fire Department purchased three of these rearmount aerials on 1992 Spartan chassis with 106-foot *LTI* ladders, built by *Quality Fire Apparatus* of Talladega, Alabama. *WMC*

Engineering had mounted its rugged *Aerialscope* telescopic platform on the Mack CF chassis exclusively since 1970. New York City (FDNY) had purchased many of these heavy-duty tower ladders over the previous two decades. But with the Mack CF no longer available, Baker had to find another custom chassis on which to mount its 75- and 95-foot *Aerialscope* booms. Some were subsequently delivered on Simon/Duplex chassis. With Mack's exit from the fire apparatus industry, New York City had switched to Seagrave as its preferred pumper and aerial chassis supplier. By the mid-1990s, Baker *Aerialscopes* were being delivered on FWD/Seagrave "J" Series custom chassis more or less exclusively. A few were delivered on Spartan chassis.

Three independent aerial ladder manufacturers were making steady inroads on the industry as the 1990s began. *Smeal Fire Apparatus* of Snyder, Nebraska, had been making its own aerial ladders and elevating platforms since the 1960s. From the mid-1980s through the early 1990s, Smeal supplied aerial devices to industry leader *Pierce Mfg.* By the time Pierce started making its own aerial ladders in 1993, Smeal Fire Apparatus had expanded into a full-line manufacturer serving a national customer base. The company's comprehensive aerial products line included 55-, 75- and 105-foot rearmount aerial ladders and 85- and 100-foot aerial platforms.

Two other independent aerial ladder makers entered the ferociously competitive aerial market in the 1990s. Formed in the late 1980s, *RK Aerials* (named for founder Rob Kreikmeier) set up shop in Fremont, Nebraska. In 1995, RK Aerials moved into a large, new plant in Fremont, and introduced its first aerial platform the following year. The company's initial OEM customers were *Central States Fire Apparatus* of Lyons, South Dakota, and Canadian fire apparatus builder *Tibotrac.* In 2000, *RK Aerials* became part of *Rosenbauer America* – the Austrian firm that had earlier bought up *Central States* and *General Safety* of Wyoming, Minnesota.

Another new independent aerial ladder maker also arrived on the scene in the early 1990s. Several former *LTI* employees formed *Aerial Innovations* in Reinholds, Pennsylvania, in 1993. This company's initial product was a heavy-duty three-section 75-foot rearmount aerial with 500-pound tip load – followed shortly by a four-section 100-foot aerial designed for rearmount, midship or tractor-drawn aerial truck applications. *AI* acquired the Boardman Company's aerial products line in 1995. Along with rival *LTI,* by the end of the decade *Aerial Innovations* had become a component of the new, revitalized *American LaFrance.*

After just 16 years in business, *Grumman Emergency Products* closed its doors in mid-1992. Not long after it was formed in 1976, Grumman purchased *Howe Fire Apparatus* – which in turn had earlier acquired the *Oren Roanoke Corp.* of Roanoke, Virginia, and *Coast Apparatus* of Martinez, California. For a time, Grumman operated Howe and Oren as separate entities but in 1980 closed the Howe plant in Anderson, Indiana, and consolidated all fire apparatus production in the former Oren plant in Roa-

1992 *LTI/Spartan* 100-foot rearmount aerial, Seattle, Washington. *Dan Jasina*

One of the world's busiest airports, *Chicago O'Hare International Airport* is protected by several Chicago fire stations. Structural fire apparatus includes this chrome-yellow 1992 E-One pumper/tanker with 2000 gpm pump, 750-gallon tank, 300-gallon foam tank and 50-foot *TeleBoom*. *WMC*

noke. Grumman's aerial products utilized the same *LTI* aerials and platforms Howe had used, but some were also delivered with Hahn *Fire Spire, Boardman Readi-Tower* and *Snorkel* aerial devices.

Grumman had introduced its own *AerialCat* 95-foot ladder-tower in 1982, followed by 102- and 121-foot versions. When Grumman threw in the towel, Kovatch Mobile Equipment (*KME*) of Nesquehoning, Pennsylvania, acquired Grumman's aerial products line, continuing production of the *AerialCat* on Grumman's low-profile *Panther* chassis in the former Oren/Grumman plant in Roanoke, Virginia.

Another significant new player was also making its presence known. Founded in Holden, Louisiana, in 1988, it didn't take long for *Ferrara Fire Apparatus* to move beyond pumpers and rescue units into the booming aerial truck market in the early 1990s. It's first such product was a 75-foot rearmount quint with RK aerial. In the late 1990s, Ferrara introduced its own *Inferno* custom chassis. Ferrara switched to Smeal aerial ladders in 1999, offering a full line of 55-, 75- and 100-foot midship and rearmount aerials and 85- and 105-foot aerial towers.

For the second time in a decade, former industry giant *American LaFrance* went out of business in 1984. Reincarnated by corporate conglomerate Figgie International and relocated from its ancestral home in Elmira, New York, to Bluefield, Virginia, in 1986, born-again American LaFrance offered a wide range of custom-chassis fire apparatus on its redesigned *Century 2000* chassis – including rearmount and midship aerial ladder trucks and quints, a second-generation pumper-mounted *Water-Chief II* 75-foot aerial, and tractor-drawn 100-foot aerials. These aerial products used a strengthened, updated version of what was basically the same three- or four-section steel aerial ladder design introduced back in 1938.

Just when it looked like American LaFrance was gone for good, a white knight rode in to save the day – and one of the industry's most iconic nameplates. In 1996, James L. Hebe – then CEO of Freightliner trucks – acquired the American LaFrance name and intellectual assets. True to his word, Hebe introduced a totally new American LaFrance *Eagle* custom fire apparatus chassis at the 1996 IAFC convention. Production began at a brand-new ALF plant in Cleveland, North Carolina, in the spring of 1997. At first, American LaFrance (like Freightliner, now part of newly-formed German-owned *DaimlerChrysler)* made only a premium custom fire apparatus chassis. Bodies (mostly pumpers) were mounted on this handsome new chassis with its unmistakable ALF lineage by various independent fire truck builders. Within a short time, however, the company was an aggressive full-line producer, building complete pumpers, rescue trucks – *and* aerials.

In addition to several other fire apparatus companies – RD Murray, 3-D Metals, Boardman and Becker – the reborn and reinvigorated ALF conglomerate also shrewdly acquired three well-known makers of aerial devices – *LTI, Aerial Innovations* and *Snorkel* – the latter for the *second* time. Introduced in 1998, the reincarnated company's first aerial product

Mukilteo, Washington's 1992 E-One 95-foot Aerial Tower with 1500 gpm pump and 200-gallon water tank. *WMC*

Quebec City, Quebec, ordered several of these extremely compact rearmount aerials with five-section, 100-foot ladders. The 1993-94 aerials were built on Spartan chassis by Thibault successor *Nova Quintech,* which took over the former Pierreville Fire Trucks plant near Pierreville, Quebec. Pierce later acquired Nova Quintech's aerial designs including the ultra-compact *SkyFive* ladder on this rig. *WMC*

was the *Silver Eagle* – a tandem-axle, 100-foot rearmount aerial with *Aerial Innovations* ladder. Before long American LaFrance's impressive product portfolio included an entire family of aerials: rearmounts, tractor-drawn aerials and a midship-mount aerial platform. Philadelphia, Pennsylvania, purchased substantial numbers of the new ALF pumpers and tractor-drawn aerials in the late 1990s.

Seagrave, meanwhile, had replaced its venerable single-axle *Rear Admiral* rearmount aerial with the *Patriot* – an all-new three- or four-section aerial that had its dual lifting cylinders under the base of the ladder. The last of the old-style steel aerials, originally introduced in 1936, were delivered in the late 1980s. Among the last of these was a white-painted tractor-drawn 100-footer for New Haven, Connecticut. In 1992 Seagrave introduced its heavy-duty *Apollo* aerial – a three-section, 105-foot aerial, and the *P-75* – a single-axle Patriot quint. A new *Meanstick* 75-foot aerial quint was added in 1997, and the *RA110* aerial was discontinued the following year. Seagrave custom apparatus was now built on the company's boxy "J" Series chassis, or new tilt-cab "T." Seagrave continued to deliver substantial numbers of pumpers, aerials and *Aerialscope* tower/ladders to the New York City (FDNY) Fire Department. In 1999, FWD/Seagrave acquired *Baker Equipment and Engineering,* which had produced the *Baker Aerialscope* since 1970. Aerialscope boom production was later moved from Baker's plant in Richmond, Virginia, to Seagrave's main plant in Clintonville, Wisconsin.

Although this highly successful, family-owned company was now also building custom-chassis pumpers, Sutphen's bread-and-butter product was still the highly regarded *Sutphen Aerial Tower.* With its distinctive "huck-bolted" lattice type box-boom and big four-door cab with prominent grille, hundreds of Sutphen 85- and 100-foot midmount aerial towers and quints were in service in fire departments across the nation. Sutphen was also making a popular *70-foot Mini-Tower Quint,* midmount straight aerial ladders and even some tillered, tractor-drawn aerial ladder trucks.

In 1998, Sutphen exhibited the *Sutphen Super Tower.* This massive tractor-trailer midship-mount Sutphen Tower featured a unique tiller cab design – the tillerman's seat was actually *inside* the crew basket. At the fire scene, the entire "cab" rose into the air, the cab roof becoming the floor of the crew basket. Only one was built.

During the 1990s, there were some notable corporate developments at *Pierce Mfg.* – the Appleton, Wisconsin, firm that now dominated the U.S. fire apparatus industry the same way American-LaFrance once had. In the early 1990s, Pierce opted out of its arrangement with Smeal and started making its *own* aerial ladders and platform/towers. Later in the decade, aerial production was moved from Appleton to Kewaunee Engineering in Kewaunee, Wisconsin. In 1996, Pierce was acquired by the much larger *Oshkosh Truck Corp.,* and two years later Pierce purchased the aerial devices line of Canadian manufacturer *Nova Quintech.* With this acquisition, Pierce got Nova Quintech's *SkyArm* and *Skypod* aerial platforms, *SkyBoom* and compact five-section *SkyFive* aerial ladder.

Pierce introduced a new 100-foot aerial platform in 1991, and new 75- and 105-foot aerials and 100- and 105-foot platforms the following year. In 1994, the company announced its high-tech *ALL-STEER* all-wheel steering system, which was optionally available on Pierce four- and six-wheel aerial apparatus. Two years later, the company came out with its unique, angular-looking *Quantum* custom chassis. With its multifaceted front end, protruding grille, retractable cab steps and side windows extending into the roof, the top-line *Pierce Quantum* chassis was also available under the company's aerials and platforms.

In 1998, Pierce finally started building tractor-drawn aerial ladder trucks. Pierce's TDAs were unique in that the aerial ladder sloped down to the rear in its travel position, optimizing forward visibility for the tillerman.

The industry's other major player – *E-One Inc.* of Ocala, Florida – also continued to field new and improved aerial products, among them all-new *CR100* and *HP75* aerials in 1995, and the *HP100* (High Performance) aerial the following year. In 1996, E-One replaced its original *Cyclone* custom chassis with a new *Cyclone II.* Also in 1996, E-One's parent company, Federal Signal, acquired Finnish aerial platform maker *Bronto Skylift.* E-One became Bronto's North American distributor, mounting *Skylift* articulating aerial platforms on its new *Cyclone II* custom chassis. The largest of these were truly monstrous 170-foot models on massive four-axle chassis.

As the 20th Century drew to a close, let's review the major milestones in aerial fire apparatus development.

1860s – First practical aerial ladder is introduced by Daniel D. Hayes.

1902 – Coil springs replace manpower in raising the aerial ladder.

1930s – Hydraulic, all-steel aerial ladders succeed spring-raised wood aerials.

1950s – Articulated elevating platforms (Snorkels) revolutionize aerial firefighting.

1960s – Telescopic aerial towers and compact rear-mount aerials introduced.

Shawnee, Oklahoma, ordered a 1993 Boardman pumper/aerial on popular four-door *Freightliner Business Class* chassis. Aerial device is a 55-foot *Apache Readi-Tower. Steven B. Loftin*

Into The Millennium

As the 21st century begins, three basic types of aerial devices are in general use by the fire services across the United States and Canada. The two most popular types are the conventional three- or four-section telescopic *aerial ladder*, and the telescopic combined *aerial ladder/elevating platform* with crew bucket. The third, less commonly encountered type, is the articulating type aerial platform, as typified by the *Bronto Skylift.*

The majority of aerial ladders in use today have prepiped waterways and remote-control master stream nozzles. Elevating platforms typically also feature integral waterways and one or two master stream nozzles in their baskets. High-tech safety features include automatic vehicle leveling and safety-override devices and TV cameras in the basket. Modern aerial devices offer wide, unobstructed walkways from turntable to tip, with high side handrails for firefighter and civilian rescue safety. Aerial ladders typically range in working height from 50-60 feet to 135 feet. While three- and four-section styles are by far the most common, some "compact" aerials have five sections and longer ladders as many as six.

Almost as common as the conventional aerial ladder now are smaller combined aerial ladder/water tower devices typified by *Snorkel's* ubiquitous *TeleSqurt* – which has been widely imitated by numerous other manufacturers. Other specialized aerial devices include: the *Squrt* – an articulated master stream boom; the *Snozzle* – a Squrt-style articulating arm with metal-piercing tip, designed for aircraft fire/rescue work; and such recent innovations as Ferrara's mighty *Strong Arm* and Crimson Fire's *Boomer* – the latter a multi-tasking water, light and power/air supply tower,

Most of the aerial apparatus in service as the first decade of the new millennium winds down are of the *rearmount* type, with the turntable at the rear of the truck and the ladder or tower boom overhanging the cab. Although in the minority, the classic *midship*-mount design – with the ladder/boom overhanging the rear of the rig – has never really gone out of style. This holds true for both conventional aerial ladders *and* aerial platforms. Similarly – although their numbers have declined sharply since their heyday around the middle of the last century – the glamorous tractor-drawn, tillered aerial ladder trucks still survive. The vast majority of aerial ladder trucks and elevating platforms in service today are equipped with pumps (quints), water tanks and preconnected attack hose lines.

Among the most popular and versatile aerial trucks in general use today – especially in smaller fire departments – is the single rear axle "junior quint," basically a standard triple-combination pumper equipped with a 65- or 75-foot rearmount aerial ladder, a configuration American LaFrance popularized with its 75-foot *Water Chief* pumper/aerial in 1979. Heavier aerials and towers invariably utilize tandem rear axles.

While the makers of aerial devices have continued to constantly improve and refine their products, there were no major new, groundbreaking developments in aerial design as the new century unfolded. But there *were* some major corporate developments, however. *Seagrave* – which in the late 1990s had acquired Baker Engineering – makers of the *Baker Aerialscope* – moved Aerialscope boom production from Richmond, Virginia, to its home plant in Clintonville, Wisconsin.

In 2002, *American LaFrance* closed its five-year-old plant in Cleveland, North Carolina, and transferred all fire apparatus production – including its *Eagle* and *Metropolitan* custom chassis – to a new facility in North Charleston, South Carolina. The company's corporate headquarters was similarly relocated to nearby Ladson, South Carolina. Five years

Perfectly posed for photographers, 1994 *Simon/Duplex LTI* 100-foot, tractor-drawn aerial in service in Seattle, Washington. Safety rules require use of wheel chocks at *all* times. *Daniel Jasina*

later, in mid-2007, American LaFrance moved into a spectacular new headquarters and plant complex in Summerville, South Carolina. Within months, however, ALF filed for bankruptcy – from which it emerged in mid-2008. While ALF custom chassis continued to be made at the Summerville plant, aerial truck production was farmed out to the former LTI plant in Ephrata, Pennsylvania, and pumper production to R.D. Murray in Hamburg, New York. The company's former *LTI* and *Aerial Innovations* operations lines had earlier been consolidated into ALF's *Aerials Division.*

E-One also emerged from a period of financial turbulence in 2008. Corporate parent *Federal Signal Corp.* of Chicago – which had owned E-One since 1979 – sold the Ocala, Florida, company to private equity firm *American Industrial Partners.*

For much of the past century, aerial ladder trucks were built only by a few of the largest fire apparatus builders – American-LaFrance, Seagrave, Pirsch, etc. Today, more than a dozen U.S. manufacturers offer a wondrous array of aerial devices to fire departments from the smallest volunteer outfits to the largest metropolitan and regional fire-rescue services.

Following is an overview of the aerial devices on the market in 2008 – a snapshot of aerial firefighting technology in America in the early 21st century – 141 years after Daniel D. Hayes introduced the first successful, mass-produced aerial ladder in the U.S.

American LaFrance LLC – Summerville, S.C. & Ephrata, Pa.
75-, 90- and 100-foot *LTI* midship-mount aerials
Metrostik 75- and 100-foot heavy-duty *LTI* rearmount aerials
100- and 105-foot tractor-drawn *LTI* tillered aerials
50-, 65- and 75-foot *TeleSqurt* aerial ladder/water towers, 55-foot *Snorkel*
75-, 93- and 100-foot *LTI* rearmount and midship aerial platforms
Chassis: *Eagle* and *Metropolitan* Custom, *Freightliner* and other heavy-duty commercial chassis.

Seagrave Fire Apparatus LLC – Clintonville, Wisconsin
Meanstick 75-foot Aerial/Quint, single rear axle
Force 100-foot Rearmount Aerial
Tractor-Drawn 100-foot Tillered Aerials
Midmount 75- and 95-foot *Aerialscope* Aerial Platforms
Chassis: Seagrave *"J"* and *"T"* Series custom chassis

Pierce Manufacturing, Inc. – Appleton, Wisconsin
100-foot midmount aerial; 75- and 105-foot Heavy-Duty aerials, 100-foot *Super Heavy-Duty* aerial, 75- and 100-foot aluminum aerials; 75- and 100-foot medium-duty aerials; 100-foot tractor-drawn tillered aerial.
95-foot midmount aerial platform
85- and 100-foot rearmount aerial platforms
Sky Arm 100-foot articulated aerial ladder/platform
Sky Boom 55- and 61-foot aerial ladder/water tower
Chassis: *Pierce Velocity, Impel, Quantum, Arrow XT* and *Saber* custom

E-One, Inc. – Ocala, Florida
Tele-Boom 50-foot midmount aerial/water tower
HP75, HP100, HM100, CR100, 110-foot rearmount aerials
Super Tiller tractor-drawn 100-foot tillered aerials
95-foot rearmount and midship-mount aerial platforms, HP100-foot rearmount platform
Chassis: *E-One Cyclone II, Typhoon Custom*

Sutphen Corporation – Amlin, Ohio
SPH 100 midship-mount aerial tower
SP 110 Magnum 110-foot midship-mount aerial tower
SP 100 midship aerial platform
SP 95 midship aerial platform
SP 90 midship aerial platform
SP 70 midship aerial platform
SL 75 and *SL104* midship-mount aerials
SA 65 and *SA75* midship-mount aerials
Chassis: *Sutphen Custom*

KME Fire Apparatus – Nesquehoning, Pa.
55- and 75-foot *Fire Stix* rearmount aerials
75-, 100- and 121-foot *AerialCat* rearmount aerials

95- and 100-foot midmount aerials
75- and 100-foot tractor-drawn tillered aerials
95- and 102-foot *AerialCat* rearmount aerial platforms
81-, 95- and 100-foot midship-mount aerial platforms

Crimson Fire Inc. – Brandon, South Dakota
75- and 103-foot rearmount aerials
100- and 103-foot tractor-drawn tillered aerials
100-foot rearmount aerial platform
100-foot midship-mount aerial platform
Boomer 28-foot water tower/lighting/electrical tower
Chassis: Spartan Custom

Ferrara Fire Apparatus – Holden, Louisiana
HD-57, HD77, HD107, HD127 rearmount aerials
HD-100 midmount aerial
HD-85 and *H-100* rearmount aerial platforms
HD-85 and *H-100* midmount aerial platforms
Strong Arm heavy-duty piercing boom (50-foot)
Chassis: *Ferrara Inferno* and *Igniter* custom

Rosenbauer America (General, Central States) – Wyoming, Minnesota and Lyons, South Dakota
Viper 60-, 65-, 75-, 85- and 104-foot rearmount aerials
Cobra 85- and 104-foot rearmount and 75-foot midship-mount aerial platforms
T-Rex 93-foot rearmount aerial platform
Raptor 102-foot rearmount aerial
Road Runner 51- and 68-foot water tower/aerial
Chassis: *Commander AT, HME 1871, Spartan, Commercial*

Smeal Fire Apparatus – Snyder, Nebraska
55-, 75-, 100-, 105- and 125-foot heavy-duty rearmount aerials
100-foot midship-mount aerial
100-foot tractor-drawn, tillered aerial
85- and 100-foot midship-mount aerial platforms
87- and 100-foot rearmount aerial platforms
Chassis: *Smeal Sirius, Spartan Customs*

Alexis Fire Equipment Co. – Alexis, Illinois
75- and 109-foot *Stick* aerials
91-, 104- and 174-foot rearmount aerial platforms

HME Fire Apparatus – Wyoming, Michigan
75-foot rearmount quint
104-foot rearmount aerial platform
109-foot rearmount aerial
Chassis: HME Custom

Various other well-known manufacturers – *Darley, Alexis, etc.* – also build aerial apparatus utilizing ladders and platforms purchased from independent OEM suppliers such as *Smeal.*

1994 Seagrave *Patriot* 100-foot, tractor-drawn aerial, Truck 7 for the Baltimore City Fire Department. *Daniel Jasina*

Baltimore City Fire Department also got this 1994 Seagrave 85-foot Snorkel. *Jack Calderone*

1994 *Becker/HME* rearmount aerial with 82-foot *LTI* ladder, Oklahoma City, Oklahoma. *Steven B. Loftin*

Anderson Engineering of Langley, B.C. delivered 44 *Bronto* aerial platforms to Canadian fire departments (and one to U.S.) between 1986-1994. Anderson exported this 170-foot *Bronto* on Simon-Duplex chassis to Guam in 1994. *Dave Stewardson*

Borough of East York, Ontario, originally purchased this *Snozzle*-equipped Freightliner pumper with *Dependable Emergency Vehicles* bodywork, which joined Toronto Fire Department roster via amalgamation. *Neil McCarten*

Quebec fire apparatus manufacturer *Nova Quintech* developed its own line of aerial devices. Kitchener, Ontario, ordered this 1996 Nova Quintech 100-foot rearmount aerial on HME chassis. Pierce later purchased NQT's aerial products line. *WMC*

After nearly six decades, Seagrave finally replaced the standard steel aerial ladder design the company had originally introduced in the mid-1930s. In the early 1990s, this venerable design was replaced by the all-new *Patriot* — a much stronger design with lifting cylinders placed *under* the ladder. Wakefield, Massachusetts, got this 1995 Seagrave Patriot rearmount with 100-foot ladder and raised-roof cab. *L. Murray Young Photo*

Big-city fire protection for a small town. Popular tourist destination, Frankenmuth, Michigan, ordered this 1997 *Pierce Lance* Aerial Tower with 2000 gpm pump and 100-foot tower. *Herb Gallaway Photo*

Reincarnated for the second time in a decade, *American LaFrance* announced totally new *Eagle* custom fire apparatus chassis in 1997. Valley Hill, South Carolina, got first of only four ALF/Aerial Innovations' *Silver Eagle* aerials with R.D. Murray bodywork. *Steven B. Loftin*

Some communities were now ordering tractor-drawn aerials equipped with full-sized pumps. Kirkland, Washington, got this 1997 *Simon/Duplex/LTI* with 1250 gpm pump on the tractor, and 100-foot LTI aerial. *Dave Stewardson*

Fast-growing, born-again *American LaFrance* acquired three independent ladder makers — *Aerial Innovations, LTI* and *Snorkel* — in the late 1990s, all eventually integrated under the *American LaFrance Aerials* name. Seattle, Washington, got this 1998 American LaFrance *Eagle* tractor-drawn aerial with *Aerial Innovations* 100-foot ladder. *Dan Jasina*

Windsor, Ontario, purchased five *Sutphen Towers* — three 95 footers and two 70-foot *Mini-Tower Quints* in 1998. Truck 1 got one of the 95 footers with 1250 gpm pump. *WMC*

El Reno, Oklahoma, got this 1999 *Sutphen Mini-Tower Quint* with 1500 gpm pump and 70-foot tower. *Steven B. Loftin*

Very few articulating-type *Snorkels* were being sold anymore, but more than a few older *Pitman Snorkel* booms were being transferred from their original chassis to new ones. Glen Ellyn, Illinois' 2000 *Snorkel 31* was built by Marion on Spartan chassis with a 75-foot *Snorkel* boom. *Matt Schumann*

There was a growing trend toward *midship-mounted* aerial platforms, with lower center of gravity and travel height. Sturgis, Michigan, got this 2000 Pierce 95-foot platform with 2000 gpm pump. *WMC*

In 1999, American LaFrance added a plainer, lower-priced *Metropolitan* custom chassis to its line. Here's a 2000 ALF *Metropolitan LTI-100QS* with LTI 100-foot ladder. *Steven B. Loftin*

American LaFrance acquired *Snorkel Fire Equipment* for the second time in the late 1990s. In addition to offering Snorkel products on its own custom and Freightliner commercial truck chassis, ALF sold some of these devices to other manufacturers. Santa Ana, California, got this 2001 Becker-built ALF Eagle Quint with 65-foot *TeleSqurt. Chuck Madderom*

Aerials in action. Pierce rearmount aerial, left, and Sutphen Tower in action at five-alarm fire in Detroit. Ladder 20's ALF/LTI sets up at right. *Dan Jasina*

American LaFrance aerials were available with choice of conventional aerial ladders or midship-mount elevating platforms. Schenectady, New York, got this 2001 ALF *Eagle 134* with 93-foot LTI platform. *Neal Van Deusen*

When Pierce finally started building tractor-drawn aerial ladder trucks, the company came out with a novel, but practical design. At rest, the aerial ladder sloped *down* to the rear, greatly improving forward visibility for the tillerman. Detroit got two of these Pierce 100-foot tillers in 2001. *WMC*

American LaFrance delivered a batch of these classy *Eagle/LTI tractor*-drawn aerials with 100-foot ladders to the Los Angeles City Fire Department. This one went to busy Truck 27 in 2002. *Chuck Madderom*

A 2002 *E-One* factory demonstrator at Michigan dealer *West Shore Services*. Note the well-protected remote-control ladder pipe and mounting for the ladder signboard —standard issue for aerial trucks since Sutphen initiated this practice in the 1960s. *WMC*

Kovatch Mobile Equipment — better known as *KME* — became one of the industry's major players in the late 1980s and 1990s. KME acquired Grumman's aerial products when Grumman went out of business. Hudson, New York, got this huge *KME AerialCat* quint in 2003. *WMC*

E-One delivered this big midmounted Aerial Tower to St. Clair, Michigan, in late 2003. *WMC*

Crimson Fire was the result of an amalgamation of *Quality Fire Apparatus* of Talladega, Alabama, and *Luverne Fire Apparatus* of Brandon, South Dakota. Spartan Motors — which had earlier acquired both of these companies — incorporated them into a new corporate entity called *Crimson.* Wayne Township, Indiana, got this Crimson 2000/400/75-foot Quint on Spartan Evolution chassis in 2003. *Dan Jasina*

2005 *Pierce Enforcer* 1500/500/75-foot quint for Burlington, Wisconsin. Note higher roof at rear of cab to accommodate standing firefighters. *Dan Jasina*

2005 American LaFrance/LTI 2000/300/100-foot quint for Buffalo, New York. Note dual nozzles in crew basket and massive ladder construction with gusset plates at top and bottom of struts. *Dan Jasina*

2006 *Ferrara Inferno Quint* with 1500 gpm pump and 77-foot aerial ladder, Elkhart Lake, Indiana. *HME* supplies proprietary *Inferno* chassis/cab to *Ferrara Fire Apparatus* in Holden, Louisiana. *Dan Jasina*

One of Europe's largest fire apparatus manufacturers, *Rosenbauer* acquired *General Safety* of Wyoming, Minnesota, and *Central States Fire Apparatus* of Lyons, South Dakota, which it amalgamated into *Rosenbauer America.* This 2006 Rosenbauer/Central States 85-foot platform with *RK* aerial (now also part of Rosenbauer America) was built for Madison, South Dakota. *WMC*

2006 *Pierce Dash* 1500/300/105-foot Quint was built for Irving, Texas, where it runs as Ladder 8. *Dan Jasina*

Troy, Michigan's chrome-yellow 2006 *Sutphen Aerial Tower* with 1500 gpm pump, 300-gallon tank and 100-foot tower with high handrails. *Dan Jasina*

One of the Toronto Fire Department's two 2006 *E-One Cyclone II Brontos.* Tower 331 has 1750 gpm pump and 114-foot *Bronto Skylift* platform. *Dave Stewardson*

The long, flexible reach of the *Bronto* is clearly illustrated by this E-One Bronto photographed at drill session at Toronto Fire Academy. Note articulated upper boom section and side-mounted escape ladder. *WMC*

A loyal American LaFrance customer, Livonia, Michigan, took delivery of this state-of-the-art ALF Eagle with 100-foot LTI tower and 2000 gpm pump in 2006. *Dan Jasina*

North Syracuse, New York's *Truck 1,* 2007 E-One 95-foot midship-mount platform with 2000 gpm pump and 95-foot tower. *Dan Jasina*

Toronto Fire Department has purchased a large number of Smeal pumpers and aerial ladder trucks, including this 1500/500/105-foot Quint delivered in 2007. Note UK-style checkerboard striping on front of cab. *Neil McCarten*

HME fire Apparatus of Wyoming, Michigan, has leased the legendary *Ahrens-Fox* name for its top-line pumpers and aerials. Owosso, Michigan, ordered this 2007 HME/Ahrens-Fox Quint with 2000 gpm pump and 104-foot *RK Aerials* ladder. *Dan Jasina*

Fire departments all over the U.S. and Canada have adopted the Chicago Fire Department's traditional black-over-red color scheme. Fraser, Michigan, got this 2007 *Pierce Dash Quint* with 2000 gpm pump and 85-foot platform. *Dan Jasina*

2007 Ferrara/Smeal 2000/400/75-foot Rearmount Quint, Summit Township, Michigan. *Dan Jasina*

Pierce introduced all-new *Pierce Velocity* and *Impel* customs in 2006. The wealthy enclave of West Bloomfield, Michigan, got this 2008 *Pierce Impel Quint* with 1500 gpm pump, 500-gallon tank and 75-foot aerial. *Dan Jasina*

America's first successful *Water Tower* was patented by John Logan and Abner Greenleaf of Baltimore, Maryland, in 1880. The first *Greenleaf Water Tower* was loaned to New York City for trial that same year and purchased by the city in 1882. The horse-drawn Greenleaf tower had a pivoting 50-foot mast. *MLC*

CHAPTER ELEVEN: THE WATER TOWER

For more than eighty years – from the 1880s until the last surviving examples were displaced almost overnight by the revolutionary new elevating platform in the late 1950s and `60s – most big-city fire departments in the United States had at least one *Water Tower* on their fire equipment rosters.

Uniquely American, the dedicated, single-purpose water tower was the heavy artillery of the fire service – the weapon of last resort when flames raged unchecked through multi-storied structures. Essentially a portable standpipe, the water tower was designed to hurl powerful streams into the blazing upper floors of America's ever-taller commercial and tenement buildings.

While other inventors had tinkered with the idea earlier, the first *successful* firefighting water tower was patented by John B. Logan and Abner Greenleaf of Baltimore Maryland, in 1880. The first Greenleaf Water Tower, with pivoting 50-foot tubular steel mast, was loaned to New York City which ultimately purchased the machine two years later. At about the same time, in 1882, the Chicago Fire Department placed a *Petrie-Ashworth* water tower into service.

But it was George C. Hale who really advanced the big-city water tower concept. Chief of the Kansas City (Missouri) Fire Department and also owner of the *Kansas City Fire Department Supply Co.,* Mr. Hale introduced a horse-drawn water tower of his own design in 1886, followed by a much-improved design with water-raised mast two years later. The base of the Hale Water Tower was at the forward end of the vehicle, with the 55-foot two-piece telescopic steel mast and curved nozzle overhanging the rear of the apparatus. Most Hale towers also had a manually operated monitor nozzle mounted on the tower deck.

More than fifty *Hale Patent* 55- and 65-foot water towers were delivered to American fire departments over the next four decades. From 1886 through 1892, George C. Hale had a virtual monopoly on water tower sales in the United States. But in 1893 a competitive challenger arrived on the scene. The Fire Extinguisher Mfg. Co. of Chicago introduced the *Champion Water Tower* designed by FEMCO's sales manager, Ernest F. Steck. Faced with his first serious competition, Hale sold his water tower patents to FEMCO in 1894 – which perpetuated the monopoly for another three years, until 1897.

Except for a few early examples with front-mounted masts, the base of the drop-frame 65-foot *Champion* water tower was at the extra-wide rear of the apparatus, with the mast nozzle extending out over the heads of the three horses needed to pull the heavy apparatus. A total of 21 *Champion* water towers were constructed over a 14-year period from 1893 to 1907.

In 1898, Henry H. Gorter, superintendent of the San Francisco Fire Department shops, introduced what would ultimately be America's tallest water tower. Three horse-drawn Gorter towers were built – two 65-footers (one each for San Francisco and Los Angeles) and a 76-foot tower for the S.F.F.D. Interestingly, the last two Gorter water towers delivered were a pair of stubby 35-footers built by the Union Machine Co. on Mack AC Bulldog chassis for San Francisco in 1927.

As was the case with most early aerial ladders, water towers were sometimes built under license by other manufacturers. American-LaFrance and Seagrave, for instance, built (and rebuilt) both Hale- and Champion-patent water towers in the late 1800s and early 1900s.

The first water towers were manually raised by firefighters furiously turning cranks or hand wheels at the tower's base. In 1893, the ever-inventive Mr. Hale devised a system by which the heavy tower mast was raised from its bed by chemical or hydraulic pressure. A large reservoir tank was mounted on the truck's frame. A chemical reaction (soda and acid mixed with water, the same used in chemical firefighting apparatus of the era) was utilized to lift the tower mast to its vertical operating position. Later, Hale water towers relied on water pressure alone.

The upper mast pipe was then cranked to its required working height. The master stream nozzle at its tip was aimed and controlled by a series of ropes operated from the ground. Some were equipped with stay poles, which provided a measure of stability against the powerful back-pressure from the tower nozzle. Called out only to the largest fires, water towers were typically pulled by two or three horses.

In 1907, Seagrave began making Hale-patent water towers equipped with its new spring-raise hoist. Rival American-LaFrance countered two years later with Hale-patent water towers fitted with its own *American Automatic* quick-raising spring hoist. As

George C. Hale designed the highly successful water tower which bore his name in 1886. The base of the Hale tower was at the front of the apparatus, with the mast and tower nozzle overhanging the rear. In 1888 Hale perfected a *water-raised* hoist for his tower. This typical *Hale Patent* water tower was built for Boston, Massachusetts, in 1903. *MLC*

Hale Water Tower with typical three-horse hitch, Boston, Massachusetts. *WMC*

with their wood aerial ladders, a pair of compressed coil springs took the muscle work out of raising the 55- or 65-foot water tower mast.

The next major development in water tower evolution was *motorization.* As early as 1911, New York City had replaced the horses under one of its Hale water towers with a Webb-Couple Gear auto tractor. Springfield, Massachusetts, motorized its Champion tower with a three-wheeled Knox-Martin tractor the same year. Other fire departments soon began to similarly motorize their formerly horse-drawn water towers with bolt-on two-wheel conversion tractors like the Christie and A&B, or four-wheel commercial type truck tractors.

American-LaFrance built its first complete, factory-built motor-driven water tower for New Orleans, Louisiana, in 1914. The 65-foot Hale tower with American Automatic spring hoist was powered by a Type 30 front-wheel-drive tractor with six-cylinder engine. Rival Seagrave delivered its first motorized water tower to St. Paul, Minnesota, in 1917. Both were of the straight-frame, four-wheel type with rear-mounted tower masts.

Thus began the golden age of the motorized water tower. During the 1920s, Seagrave delivered three big 65-foot water towers to Chicago – in 1923, 1925 and 1928. American-LaFrance delivered seven Type 31 front-drive water towers between 1919 and 1923, and supplied Type 31 conversion tractors to motorize nine others. New York City and Albany, New York, Philadelphia, Pennsylvania, and Providence, Rhode Island, purchased complete American-LaFrance/Hale water towers drawn by Type 17 four-wheel tractors. Rochester and Utica, New York, and Indianapolis, Indiana, purchased tractor-drawn Type 117 water towers in the late 1920s. And Boston purchased a pair of ALF Type 117 tractors for two of its old water towers.

Of special note were the four Type 31 *Deluge Towers* American-LaFrance built in the 1920s. The first of these was actually a 1923 Type 31 motor tractor conversion of a formerly horse-drawn 1901 FEMCO high-pressure hose wagon with manually raised 32-foot deluge mast for Pittsburgh, Pennsylvania. In 1927, the Elmira, New York, factory built two more of these front-drive Combination Hose and Deluge Wagons for the Pittsburgh Bureau of Fire. One more example – the last – was built for Atlantic City, New Jersey, in 1929.

The Seagrave Corporation of Columbus, Ohio, built ten motorized water towers between 1917 and 1932. All but one – a tractor-drawn *Duplex* tower with dual mast nozzles for New York City – were of the four-wheel, rear mast type. The last Seagrave Water Tower went to Washington, D.C. in 1932.

The legendary big-city water tower era ended with a real flourish in the 1930s. American-LaFrance deliv-

Hale Patent water towers were built under license by both American-LaFrance and Seagrave. Cincinnati, Ohio, ordered this Hale 65-foot tower from American-LaFrance in 1905. Hale towers were available with either water- or chemical-raised hoist. *MLC*

ered four truly awesome tractor-drawn water towers at the depths of the Great Depression. They included New York City's last water tower – a 65-foot *Duplex* delivered in 1930 – and a massive white-painted 1930 *Master Series* 65-foot water tower with rear-mounted mast for Jersey City, New Jersey. In 1936, American-LaFrance built what was purportedly the largest and most powerful water tower ever built in America – a stunning white-painted 65-foot American Automatic with rear-mounted mast and classic 400 Series tractor. The last water tower built in America was a one-off Combination Water Tower and City Service Hook and Ladder truck with closed-cab tractor shipped to Los Angeles, California, in 1938.

In the meantime, many big-city fire departments – Boston and Baltimore among them – continued to regularly rebuild and modernize their venerable water towers – some of which were originally purchased in the 1880s and 1890s. Because they were seldom used – sometimes years between big fires – water towers were often relegated to the side bays of downtown fire stations and manned part-time or only when needed.

Since the 1930s, much more versatile hydraulically-operated, steel aerial ladders with fixed or demountable ladder pipes were increasingly rendering the traditional single-purpose water tower obsolete. In 1948 the New York City Fire Department (FDNY) placed its first three metal aerial ladder trucks into service – one each from American-LaFrance, Seagrave and Pirsch – ultimately sealing the fate of the city's six water towers. All had been retired by the late 1950s.

Yet, look at any photo taken at major fires in America's largest cities in the 1940s and 1950s and you're likely to see a water tower in the foreground delivering its powerful master stream punch. No Chicago "5-11" was complete without at least one of these veteran iron dinosaurs visible through the smoke.

Ironically, it was in the Windy City that the classic American water tower ultimately met its Waterloo. In 1958, the Chicago Fire Department fielded the nation's first elevating platform. Fire Commissioner Robert Quinn's experimental *Snorkel* offered his firefighters heretofore-unknown aerial mobility. Safely riding in a bucket at the end of an articulating upper arm, firefighters could reach over and beyond walls and parapets to direct a powerful master stream with unerring accuracy into the very heart of the fire. Within two years, Chicago had purchased four more Snorkels – and put all three of its faithful old water towers out to pasture.

The *Fire Extinguisher Mfg. Co.* of Chicago introduced the competing FEMCO *Champion* water tower in 1893. Most *Champion* towers had rear-mounted masts with the nozzle extending out front, above the horses. Here's a head-on view of a *Champion Water Tower* in the early 1900s. *MLC*

The elevating platform swept the fire service like a firestorm. Unlike the single-purpose water tower, the elevating platform could also perform rescue duty as well. Tradition is difficult to overturn, however. Although they quickly embraced them, some cities – like Baltimore – actually called their new elevating platforms Water Towers. By the 1970s, the few remaining water towers in America had become museum pieces. Others, unfortunately, were simply scrapped.

Water tower historian Bill Hass – author of *The American Water Towers* – estimates that some 110 water towers were built in America in the 58 years between 1880 and 1938. Less than a dozen remain.

Yet the water tower name lives on – in modern combined aerial ladder/master stream devices like today's *TeleSqurt* and *TeleBoom.*

The *Champion Water Tower* had a distinctive drop-frame chassis. All but the first few *Champion* towers had rear-mounted masts, like this one. The 55- or 65-foot mast was hand-cranked into its vertical operating position. *MLC*

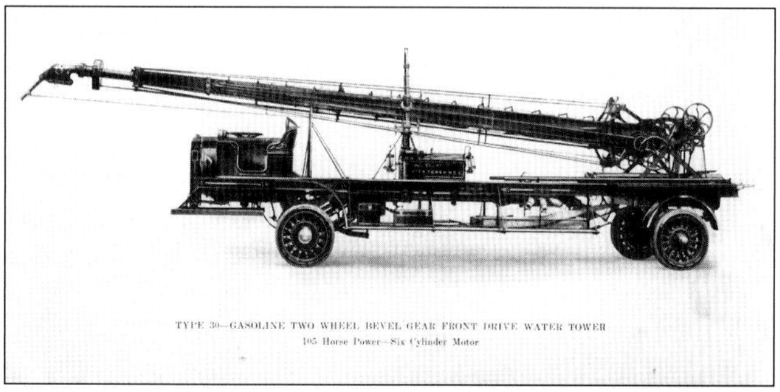

American-LaFrance built its first motorized water tower for New Orleans, Louisiana, in 1914. The *Type 30* tower was powered by a two-wheel front-drive tractor with 105 horsepower six-cylinder engine and had a spring-raised 65-foot *Hale Patent* mast. *WMC*

Most large cities motorized their formerly horse-drawn water towers in the teens. New York City purchased four Webb/Couple-Gear electric tractors in 1911 and 1912. The 1912 Webb shown here in a parade was attached to Water Tower 3's Hale tower.

Seagrave delivered its first motorized water tower to St. Paul, Minnesota, in 1917. Amazingly, more than 90 years later, it's still on the road! Powered by a Seagrave six-cylinder engine, St. Paul's big tower has a rear-mounted, 65-foot spring-raised mast. Note the twin windshields and tall, midship-mounted deck turret. *MLC*

The Chicago Fire Department purchased three identical Seagrave 65-foot water towers — in 1923, 1925 and 1928. Two of them are still around today. These big towers were a fixture at every major fire in the Windy City — until the newfangled elevating platform put them all out to pasture in the late 1950s. *MLC*

Superintendent of the San Francisco municipal shops, *Henry H. Gorter* designed yet another style of water tower in 1898. Three of these massive Gorter water towers — two with 65-foot masts and one 76 footer, the tallest in the U.S., were built in the late 1800s and early 1900s. The San Francisco Fire Department used this 65-foot Gorter tower, seen here drawn by a 1926 Kleiber tractor, into the 1970s. *Bob Allen*

American-LaFrance built more motor-driven water towers than any other fire apparatus manufacturer. This 1923 *American Automatic* water tower drawn by a *Type 17* four-wheel tractor and with tillered rear steering and rear-mounted 65-foot mast was built for Providence, Rhode Island. Note the two large turret nozzles on the trailer deck. *MLC*

American-LaFrance water towers were built in two basic styles — the tractor-drawn *Type 17* with four-wheel tractor and trailer, or the straight-frame *Type 31* with two-wheel front-drive motor tractor. Syracuse, New York, and Dallas and Houston, Texas, all purchased 65-foot American-LaFrance *Type 31* water towers in 1923. The six-cylinder Type 31 tractor on the Syracuse tower shown here was replaced with a four-wheel Seagrave aerial truck tractor in the 1950s. *WMC*

Water tower production in America peaked in the 1920s. Albany, New York, got this American-LaFrance *Type 17 American Automatic* 65-foot Water Tower with midship-mounted mast in 1926. Still in complete, virtually original condition today, it's on public view in the New York State Museum in Albany. *MLC*

Rochester and Utica, New York, and Indianapolis, Indiana, all purchased American-LaFrance tractor-drawn water towers with three-section 65-foot masts and *Type 117* tractors in 1927-28. Only Indy's survives, in the care of Indianapolis Fire Buffs. Still rolling on its original hard-rubber tires, Indy's 1927 ALF Hale Patent water tower is a popular participant at SPAAMFAA antique fire apparatus shows and musters in the Indianapolis area. *WMC*

The last two *Gorter Patent* water towers were constructed for the San Francisco Fire Department by the Union Machine Co. on Mack AC "Bulldog" chassis in 1927. These stubby towers on hard-rubber tires with 35-foot masts remained in service into the 1970s. Note the distinctive ball-socket *Gorter* nozzle on the tower's tip. *WMC*

American-LaFrance built four of these special *Type 31 Deluge Wagons* — three for Pittsburgh, Pennsylvania, and one for Atlantic City, New Jersey. This is one of the two delivered to the Pittsburgh Bureau of Fire in 1927. All had high-pressure hose bodies and manually raised 32-foot water tower masts. *MLC*

One of New York City's *Hale* water towers rises into the air at extra-alarm fire in the 1920s. Formerly horse-drawn tower has been motorized with a Mack AC tractor. Note the Mack AC wood aerial on the left, the powerful high-pressure hose wagon stream — and the billowing smoke from a steam fire engine hard at work in the background! *MLC*

Not surprisingly, the New York City Fire Department operated more water towers — six in all — than any other city. At least one of these towers attended every extra-alarm fire in the city. Powered by a 1930s-vintage *Walter* tractor, here's one of the FDNY's towers in action at a woodworking mill fire on Water St. in 1939. *MLC*

Last of the more than 300 American-LaFrance *Type 31* aerial trucks and water towers built since 1915 was this 1929 *Type 31-4 Deluge Tower* built for Atlantic City, New Jersey. Identical to the two delivered to Pittsburgh, Pennsylvania, two years earlier, the Atlantic City tower has a 32-foot mast and two fixed deck pipes. In a private museum collection today, Reg. #6615 is still occasionally taken out for antique fire apparatus shows and musters. *Dan G. Martin Photo*

New York City continued to purchase water towers into the 1930s. The city got its last new towers in 1929 and 1930. The only tractor-drawn water tower Seagrave ever built, this impressive 65-foot *Duplex Water Tower*, was delivered in early 1929. As the name implies, the *Duplex* tower had two mast nozzles — one at the top of the base section and the other atop the telescopic upper section. *MLC*

Rear view of New York City's 1929 Seagrave *Duplex* Water Tower. Note the two big-bore mast nozzles and fixed deck turrets, hose inlets under the trailer deck, and the low-set tiller seat. The tillerman's seat was offset slightly to the left, so he could see around the midship-mounted tower mast. *MLC*

The FDNY got its last water tower — a tractor-drawn *Type 117* American-LaFrance — in 1930. Here, the new tower is being demonstrated for Gotham taxpayers shortly before it went into service. Note the *Duplex* tower nozzles — one each on upper and lower sections of 65-foot mast; two big deck turrets and tiller seat overhanging the rear end of this firefighting battlewagon. *WMC*

Two water towers in action at a SPAAMFAA Muster in St. Paul, Minnesota, 1994. At left is St. Paul's 1917 Seagrave, in background Fargo, North Dakota's 1920/1909/1951 American-LaFrance. *WMC*

The 1930s marked the golden age — and passing — of the American water tower. American-LaFrance delivered three magnificent big-city water towers in the 1930s. Seagrave delivered its tenth, and last, motor-driven tower to Washington, D.C. in 1932. American-LaFrance built this stunning *Master Series American Automatic* 65-foot water tower for Jersey City, New Jersey, in 1930. *MLC*

Said to be the most powerful water tower ever built in America, this truly awesome, white-painted 65-foot *American Automatic* water tower with American-LaFrance 400 Series tractor was built for Newark, New Jersey, in 1936. Words fail us, but the picture says it all. Sadly, this beauty was *driven* to the junkyard 22 years later. *WMC*

The last new, factory-built water tower made in America — and the only one ever built with a closed cab — was this one-of-a-kind *Combination Water Tower and City Service Ladder Truck* built for Los Angeles, California, in 1938. The ladder racks were later removed and the 400 Series American-LaFrance was converted into a straight water tower. Note the tillerman's seat tucked under the 65-foot tower mast. *WMC*

Hoist detail of American-LaFrance *Hale Patent* Type 17 Tractor-Drawn Water Tower delivered to New York City in 1923. Note water-raise reservoir on trailer bed, hose from inlets to base of tower mast, and heavy steel stabilizer pole attached to mast pivot. *MLC*

Not all of the water towers built went to big cities. Some smaller cities boasted towers, too. Louisville, Kentucky, bought a horse-drawn *Hale* 55-footer away back in 1892. The last in a succession of auto tractors placed under it was this 1942 Chevrolet. After nearly 120 years it's still on the Louisville Fire Department roster – albeit as part of the L.F.D. Historical Collection. *WMC*

APPENDIX

Aerial Ladder Truck Manufacturers

American LaFrance Foamite Corp.– Elmira, New York	1882 – Current
Ahrens-Fox Fire Engine Co. – Cincinnati, Ohio	1923 – 1940
Crown Coach Corp. – Los Angeles, California	1960s – 1985
Seagrave Corp. – Columbus, OH/Clintonville, Wisconsin	1902 – Current
Peter Pirsch & Sons Co. – Kenosha, Wisconsin	1900 – 1990
Maxim Motor Company – Middleboro, Massachusetts	1939 – 1989
Mack Trucks – Allentown, Pennsylvania	1920s – 1980s
Oren-Roanoke Corp. – Roanoke, Virginia	1960s – 1980
Sutphen Corporation – Amlin, Ohio	1980s – Current
American Fire Apparatus – Battle Creek, Michigan	1950s – 1986
Hahn Motor Co. – Hamburg, Pennsylvania	1960s – 1990
Howe Fire Apparatus Co. – Anderson, Indiana	1950s – 1970s
Ferrara Fire Apparatus – Holden, Louisiana	19979 – Current

Elevating Platforms – Articulated

Snorkel Fire Equipment Co. – Grandview, St. Joseph, Missouri	1958 – Current
Mobile Aerial Towers (*Hi-Ranger*) – Fort Wayne, Indiana	1959 – 1989
Calavar Corp. (*Firebird*) – Santa Fe Springs, California	1969 – 1979
Bronto Skylift – Tampere, Finland	1966 – Current
Simon Aerials – Milwaukee, Wisconsin	1980s – 1990s
Schwing America, Inc. – St. Paul, Minnesota	1990s

Aerial Towers – Telescopic

Sutphen Corporation – Amlin, Ohio (*Sutphen Aerial Tower*)	1963 – Current
Snorkel Fire Equipment Co. (*Tele-Squrt*)	1970 – Current
Aerialscope (Truco, Baker, Seagrave)	1964 – Current
Reading Techmatic (*Readi-Tower*) – Reading, Pennsylvania	1970s – 1990s
Grumman Emergency Products (*AerialCat*) - Roanoke, Virginia	1987 – 1992

Independent Aerial Ladder Manufacturers

Combination Ladder Co. - Providence, Rhode Island	1900s-1930s
Grove Mfg. Co. – Shady Grove, Pennsylvania	1959 – 1973
Ladder Towers Inc. (LTI) – Ephrata, Pennsylvania	1973 – 2000 (ALF)
RK Aerials – Fremont, Nebraska	1988 – 2000 (ALF)
Aerial Innovations – Reinholds, Pennsylvania	1997 –
Nova Quintech – Terrebonne, Quebec	1990 – 1997
Smeal Mfg. Co. – Snyder, Nebraska	1972 – Current
Steeldraulics, Inc. – Lancaster, Pennsylvania	1980 – 1987
Sponco – Oklahoma City, Oklahoma	1950s
Memco Aerial Ladder Co. – Oklahoma City, Oklahoma	1950s

1902 Hayes 65-foot horse-drawn aerial, ex-Syracuse and Fulton, New York. *American Museum Of Fire Fighting*, Hudson, New York. *WMC*

1913 *Webb/Couple Gear 75-foot Aerial,* Medicine Hat, Alberta, Canada. Restored by the Medicine Hat Vintage Vehicle Club. *Garry Meidinger*

1913 American-LaFrance *Type 17-6* tractor-drawn 75-foot aerial still owned by Wildwood, New Jersey. *WMC*

1917 Seagrave 65-foot water tower, owned and restored by the St. Paul, Minnesota, Fire Department. *WMC*

1924 Ahrens-Fox 75-foot tractor-drawn aerial, ex-Kokomo, Indiana. Owned by Mike Adams, Burt, Michigan. *Dan Jasina*

1926 American-LaFrance *Type 31-6* front-drive 65-foot aerial, ex-West New York, New Jersey. *WMC*

1927 *Magirus* 85-foot aerial, ex-Gary, Indiana. Owned by Rodger Birchfield, Indianapolis, Indiana. *WMC*

1927 American-LaFrance *Type 117 Tractor-Drawn 65-foot Water Tower*, Indianapolis, Indiana. *WMC*

1930 Seagrave 85-foot tractor-drawn, spring-raised wood aerial, ex-Syracuse, New York. Owned by Francis Glenn, Blenheim, Ontario. *WMC*

1931 American-LaFrance *Type 231 Front-Drive 75-foot Aerial,* ex-Rapid City, South Dakota. Owned by Mike Adams, Burt, Michigan. *WMC*

1935 Mack *Model BQ* 75-foot tractor-drawn aerial with engine-driven hoist, ex-Bradford, Pennsylvania. Owned by Andy Henderson, Weedsport, New York. *WMC*

1937 *Peter Pirsch & Sons* 85-foot tractor-drawn aerial with hydro-mechanical hoist, ex-St. Paul, Minnesota. *WMC*

1938 Seagrave *Model J-12* 75-foot tractor-drawn aerial, ex-Rome, New York. Owned by Matthew Lee, Plymouth, Michigan. *WMC*

1938 *Diamond-T with Pirsch Junior 50-Foot Aerial,* ex-Leavenworth, Kansas, and York, Nebraska. *Chuck Madderom Photo*

1942 *Pirsch Intermediate* 65-foot canopy cab aerial, Chilton, Wisconsin, F.D. *WMC*

1946 American-LaFrance Model *M-5-JOX* 85-foot service aerial, ex-Rensselaer, New York. *American Museum of Fire Fighting,* Hudson, New York. *WMC*

1949 Mack *Model L-85* service aerial, Maxim 75-foot ladder, ex-Niles, Michigan. Owned by George German, Paris, Ontario. *WMC*

1950 American-LaFrance *700 Series Rescue Model* 85-foot tractor-drawn aerial, ex-Hewitt, Long Island, New York. Mahlon Irish Jr., Homer, New York. *WMC*

1951 Seagrave J-12 canopy cab tractor-trailer 100-foot aerial, ex-Canton, Ohio. *WMC*

1956 Maxim 65-foot service aerial, ex-Bucyrus, Ohio. Owned by Tom Musselman. *WMC*

1957 Mack "B"/Maxim 85-foot tractor-drawn aerial, ex-Mount Joy, Pennsylvania. Owned by Jan Rausch, Clarkston, Michigan. *WMC*

1960 Ford/Hi-Ranger 75-foot snorkel, Chicago Fire Department Snorkel #5. *Ron Heal Photo*

1960 Erlander/Ford 85-foot *Strato-Tower* snorkel, Chicago Fire Department. *Ron Heal Photo*

1962 Maxim *"S"* 85-foot tractor-trailer aerial at 2006 SPAAMFAA National Muster, Columbus, Ohio. *WMC*

1963 Mack *C-85* tractor-drawn 100-foot aerial, ex-East Rockaway, Long Island, New York, and Lapeer, Michigan. Owned by Tom Reed, Lapeer, Michigan. *WMC*

1965 Seagrave *70th Anniversary Series 900-T* 100-foot tractor-drawn aerial, ex-Lansing, Michigan. Owned by Matthew Lee, Plymouth, Michigan. *WMC*

1965 *Ward LaFrance Ambassador/Hi-Ranger 85-foot Snorkel,* ex-Smyrna, New York. Owned by Gene Billings. *WMC*

1965 American-LaFrance *900 Series* 1000 gpm/100-foot quintuple combination, ex-Steubenville, Ohio. Owned by George Lasecki. *WMC*

1966 American-LaFrance *Aero-Chief 70* elevating platform, ex-Manville, New Jersey. Owned by Mahlon Irish Jr., Homer, New York. *WMC*

1967 *FWD/Pierce/Pitman 75-foot Snorkel,* ex-Davenport, Iowa. *Steven B. Loftin*

1968 *Pirsch Intermediate* 75-foot service aerial, ex-New Ulm, Minnesota. *WMC*

1970 *American Fire Apparatus/Hendrickson/Grove* 1250/150/85-foot aerial tower, Burnsville, Minnesota. *Dave Stewardson*

1973 Mack CF-600 75-foot *Baker Aerialscope,* Leroy, New York. *WMC*

1974 *Calavar Firebird/ Hendrickson 150-foot Aerial Platform,* Sacramento, California, ex-Anchorage, Alaska. *Chuck Madderom*

1976 *Sutphen Aerial Tower,* 1000/500/85-foot, Bismarck, North Dakota. *Dave Stewardson*

1976 *Crown Firecoach/ Maxim* 100-foot tractor-drawn aerial, ex-San Diego, California. Owned by Mike McDonald, Escondido, California. *WMC*

1977 Maxim *Model "F"* 100-foot rear-mount aerial, ex-Saco, Maine. *WMC*

1978 Seagrave *Rear Admiral* 100-foot rear-mount aerial, Chicago Fire Department. *Dan Jasina*

1987 *Emergency One Hurricane 110-Foot Aerial,* East Rochester, New York. *WMC*

1996 *Pierce Quantum* 100-foot aerial quint, Bellevue, Nebraska. *Steven B. Loftin*

1996 *Simon/Duplex/LTI* 100-foot tractor-drawn aerial, Los Angeles City, California. *Chuck Madderom*

1998 *E-One Hurricane/ Bronto Skylift HLA* 167-foot aerial platform, Calgary, Alberta. *Dave Stewardson*

2000 *E-One Cyclone II* 100-foot tractor-trailer aerial, Kansas City, Missouri. *Harold R. McMann*

2001 *Pierce Dash* 100-foot tractor-drawn aerial, Minneapolis, Minnesota. *WMC*

2003 *American LaFrance Eagle/ LTI* 1500/300/100-foot tractor-trailer quint, Woodinville, Washington. *Dave Stewardson*

2007 *Smeal/Spartan* 100-foot aerial, Toronto, Ontario. *Neil McCarten Photo*

2008 *Pierce Dash* 100-foot rearmount aerial, Detroit, Michigan. *Daniel Jasina*

2008 *Rosenbauer/General 1750/250/102-Foot T-Rex Metz Aerial Platform,* Cochrane, Alberta. *Terry Yip*

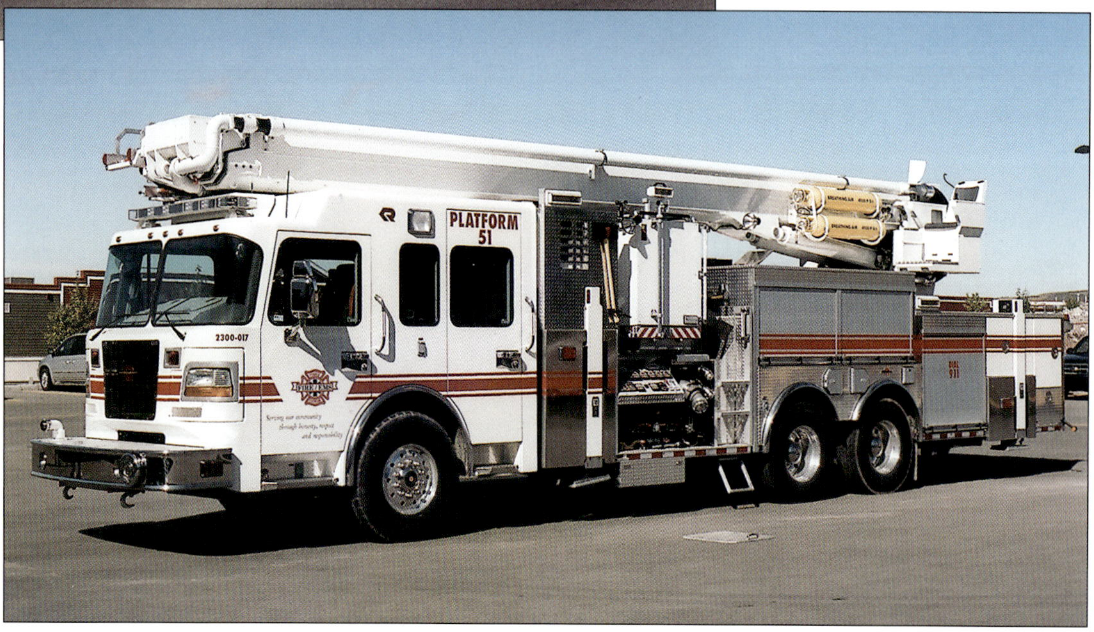

EMERGENCY VEHICLES

100 Years of American LaFrance: An Illustrated History .... ISBN 1-58388-139-5
Ahrens-Fox Fire Apparatus Photo Archive .... ISBN 1-58388-217-0
American Fire Apparatus Co. 1922-1993 Photo Archive .... ISBN 1-58388-131-X
American LaFrance 400 Series Photo Archive .... ISBN 1-58388-185-9
American LaFrance 500/600 Series: Photo Archive .... ISBN 1-58388-210-3
American LaFrance 700 Series 1945-1952 Photo Archive .... ISBN 1-882256-90-5
American LaFrance 700 Series 1945-1952 Photo Archive Volume 2 .... ISBN 1-58388-025-9
American LaFrance 700 & 800 Series 1953-1958 Photo Archive .... ISBN 1-882256-91-3
American LaFrance 900 Series 1958-1964 Photo Archive .... ISBN 1-58388-002-X
Classic Seagrave 1935-1951 Photo Archive .... ISBN 1-58388-034-8
Crown Firecoach 1951-1985 Photo Archive .... ISBN 1-58388-047-X
Emergency One Fire Apparatus 1974-2007 .... ISBN 1-58388-225-1
Elevating Platforms: A Fire Apparatus Photo Gallery .... ISBN 1-58388-164-6
Encyclopedia of Canadian Fire Apparatus .... ISBN 1-58388-119-0
Evolution of Aerial Fire Apparatus: An Illustrated History .... ISBN 1-58388-235-9
Fire Chief Cars 1900-1997 Photo Album .... ISBN 1-882256-87-5
Firefighting Tanker Trucks and Tenders: A Fire Apparatus Photo Gallery .... ISBN 1-58388-138-7
Funeral Service & Flower Cars Photo Archive .... ISBN 1-58388-227-8
FWD Fire Trucks 1914-1963 Photo Archive .... ISBN 1-58388-156-5
Grumman Fire Apparatus 1976-1992 Photo Archive .... ISBN 1-58388-165-4
Hahn Fire Apparatus 1923-1990 Photo Archive .... ISBN 1-58388-077-1
Heavy Rescue Trucks 1931-2000 Photo Gallery .... ISBN 1-58388-045-3
Imperial Fire Apparatus 1969-1976 Photo Archive .... ISBN 1-58388-091-7
Industrial and Private Fire Apparatus 1925-2001 Photo Archive .... ISBN 1-58388-049-6
John Bean and FMC Fire Apparatus .... ISBN 1-58388-237-5
*Mack Fire Trucks 1911-2005 Illustrated History** .... ISBN 1-58388-157-3
Maxim Fire Apparatus 1914-1989 Photo Archive .... ISBN 1-58388-050-X
Maxim Fire Apparatus Photo History .... ISBN 1-58388-111-5
Navy & Marine Corps Fire Apparatus 1836 -2000 Photo Gallery .... ISBN 1-58388-031-3
Pierre Thibault Ltd. Fire Apparatus 1918-1990 Photo Archive .... ISBN 1-58388-074-7
Pierce Fire Apparatus 1939-2006 Illustrated History .... ISBN 1-58388-189-1
Pirsch Fire Apparatus 1890-1991 Photo Archive .... ISBN 1-58388-082-8
Police Cars: Restoring, Collecting & Showing America's Finest Sedans .... ISBN 1-58388-046-1
Saulsbury Fire Rescue Apparatus 1956-2003 Photo Archive .... ISBN 1-58388-106-9
Seagrave 70th Anniversary Series Photo Archive .... ISBN 1-58388-001-1
Seagrave Fire Apparatus 1959-2004 Photo Archive .... ISBN 1-58388-132-8
TASC Fire Apparatus 1946-1985 Photo Archive .... ISBN 1-58388-065-8
The Illustrated Encyclopedia of American Fire Engine Manufacturers .... ISBN 1-58388-203-0
Van Pelt Fire Apparatus 1925-1987 Photo Archive .... ISBN 1-58388-143-3
Volunteer & Rural Fire Apparatus Photo Gallery .... ISBN 1-58388-005-4
W.S. Darley & Co. Fire Apparatus 1908-2000 Photo Archive .... ISBN 1-58388-061-5
Wildland Fire Apparatus 1940-2001 Photo Gallery .... ISBN 1-58388-056-9
Young Fire Equipment 1932-1991 Photo Archive .... ISBN 1-58388-015-1

RAILWAYS

American Passenger Trains: WWII to Amtrak .... ISBN 1-58388-232-4
Chicago & North Western Passenger Trains of the 400 Fleet Photo Archive .... ISBN 1-58388-159-X
Chicago Stations & Trains Photo Archive .... ISBN 1-58388-216-2
Chicago, St. Paul, Minneapolis & Omaha Railway 1880-1940 Photo Archive .... ISBN 1-882256-67-0
Chicagoland Commuter Railroads: Metra & Northern Indiana Commuter Transportation District .... ISBN 1-58388-190-5
Classic Sreamliners Photo Archive: The Trains and the Designers .... ISBN 1-58388-144-x
Freight Trains of the Upper Mississippi River Photo Archive .... ISBN 1-58388-136-0
Great Lakes Ore Docks and Ore Cars .... ISBN 1-58388-202-2
Great Northern Railway Ore Docks of Lake Superior Photo Archive .... ISBN 1-58388-073-9
Illinois Central Railroad 1854-1960 Photo Archive .... ISBN 1-58388-063-1
Interurban Trains to Chicago Photo Archive .... ISBN 1-58388-199-9
Locomotives of the Upper Midwest Photo Archive: Diesel Power in the 1960s and 1970s .... ISBN 1-58388-113-1
Milwaukee Road 1850-1960 Photo Archive .... ISBN 1-882256-61-1
Milwaukee Road Depots 1856-1954 Photo Archive .... ISBN 1-58388-040-2
Northern Pacific Railway Photo Archive .... ISBN 1-58388-186-7
Pacific Coast Commuter Railroads: From San Diego to Anchorage .... ISBN 1-58388-221-9
Pennsylvania Railroad Locomotives: Steam, Diesel & Electric Photo Archive .... ISBN 1-58388-228-6
Show Trains of the 20th Century .... ISBN 1-58388-030-5
Soo Line 1975-1992 Photo Archive .... ISBN 1-882256-68-9
Steam Locomotives of the B&O Railroad Photo Archive .... ISBN 1-58388-095-X
Trains of the Circus 1872-1956 .... ISBN 1-58388-024-0
Trains of the Upper Midwest Photo Archive Steam & Diesel in the 1950s & 1960s .... ISBN 1-58388-036-4
Union Pacific Railroad- Passenger Trains of the City Fleet Photo Archive .... ISBN 1-58388-236-7

TRUCKS

AM General: Hummers, Mutts, Buses & Postal Jeeps .... ISBN 1-58388-135-2
Autocar Trucks of the 1950s At Work .... ISBN 1-58388-231-6
Autocar Trucks 1899-1950 Photo Archive .... ISBN 1-58388-115-8
*Brockway Trucks 1948-1961 Photo Archive** .... ISBN 1-882256-55-7
Chevrolet El Camino Photo History Incl. GMC Sprint & Caballero .... ISBN 1-58388-044-5
Circus and Carnival Trucks 1923-2000 Photo Archive .... ISBN 1-58388-048-8
Diamond T Trucks 1911-1966 Photo Archive .... ISBN 1-58388-204-9
Dodge C-Series Trucks Restorer's & Collector's Reference Guide and History .... ISBN 1-58388-140-9
Dodge Heavy-Duty Trucks 1928-1975 .... ISBN 1-58388-194-8
Dodge Pickups 1939-1978 Photo Album .... ISBN 1-882256-82-4
Dodge Ram Trucks 1994-2001 Photo History .... ISBN 1-58388-051-8
Dodge Trucks 1948-1960 Photo Archive .... ISBN 1-882256-37-9
El Camino by Chevrolet .... ISBN 1-58388-215-4
Federal Trucks Photo Archive .... ISBN 1-58388-223-5
Ford Heavy-Duty Trucks 1948-1998 Photo History .... ISBN 1-58388-043-7
Ford Medium-Duty Trucks 1917-1998 Photo History .... ISBN 1-58388-162-X
Ford Ranchero 1957-1979 Photo History .... ISBN 1-58388-126-3
Freightliner Trucks 1937-1981 Photo Archive .... ISBN 1-58388-090-9
FWD Trucks 1910-1974 Photo Archive .... ISBN 1-58388-142-5
GMC Heavy-Duty Trucks 1927-1987 .... ISBN 1-58388-125-5
GMC Light-Duty Trucks .... ISBN 1-58388-191-3
International Heavy Trucks of the 1950s: At Work .... ISBN 1-58388-160-3
International Heavy Trucks of the 1960s: At Work .... ISBN 1-58388-161-1
Jeep 1941-2000 Photo Archive .... ISBN 1-58388-021-6
Jeep Prototypes & Concept Vehicles Photo Archive .... ISBN 1-58388-033-X
Kenworth Trucks 1950-1979 At Work .... ISBN 1-58388-147-6
Land Rover: The Incomparable 4x4 from Series 1 to Defender, Ludvigsen Library Series .... ISBN 1-58388-179-4
*Mack Model AB Photo Archive** .... ISBN 1-882256-18-2
*Mack AP Super-Duty Trucks 1926-1938 Photo Archive** .... ISBN 1-882256-54-9
*Mack Model B 1953-1966 Volume 2 Photo Archive** .... ISBN 1-882256-34-4
*Mack EB-EC-ED-EE-EF-EG-DE 1936-1951 Photo Archive** .... ISBN 1-882256-29-8
*Mack FC-FCSW-NW 1936-1947 Photo Archive** .... ISBN 1-882256-28-X
*Mack LF-LH-LJ-LM-LT 1940-1956 Photo Archive** .... ISBN 1-882256-38-7
Muscle Trucks: High-Performance Pickups .... ISBN 1-58388-197-2
Peterbilt Trucks 1939-1979 At Work .... ISBN 1-58388-152-2
Refuse Trucks Photo Archive .... ISBN 1-58388-042-9
Reo Trucks 1910-1966 Photo Archive .... ISBN 1-58388-181-6
Semi Trucks of the 1950s: A Photo Gallery .... ISBN 1-58388-187-5
Studebaker Trucks 1927-1940 Photo Archive .... ISBN 1-882256-40-9
The Long Haul: American Trucking Companies .... ISBN 1-58388-211-1
White Trucks of the 1950s At Work .... ISBN 1-58388-230-8
White Trucks 1900-1937 Photo Archive .... ISBN 1-882256-80-8

RECREATIONAL VEHICLES & OTHER

Commercial Ships on the Great Lakes Photo Gallery .... ISBN 1-58388-153-0
Drag Boats of the 1960s Photo Archive .... ISBN 1-58388-222-7
Great Lakes Shipping: Ports and Cargoes .... ISBN 1-58388-238-3
Illustrated Guide to Snowmobile Racing .... ISBN 1-58388-170-0
Phillips 66 1945-1954 Photo Archive .... ISBN 1-882256-42-5
RV & Camper Toys: The History of RVing in Miniature .... ISBN 1-58388-207-3
RVs & Campers 1900-2000: An Illustrated History .... ISBN 1-58388-064-X
Ski-Doo Racing Sleds 1960-2003 Photo Archive .... ISBN 1-58388-105-0
The Collector's Guide to Ski-Doo Snowmobiles .... ISBN 1-58388-133-6
Tugboats of the East Coast: A Photo Gallery .... ISBN 1-58388-224-3
Tugboats of the Great Lakes .... ISBN 1-58388-192-1
Vintage Snowmobila: A Guide to Snowmobile Collectibles .... ISBN 1-58388-169-7
Vintage Snowmobiles: The Golden Years 1968-1982 .... ISBN 1-58388-193-X

More Great Titles From
Iconografix

All Iconografix books are available from direct mail specialty book dealers and bookstores worldwide, or can be ordered from the publisher. For book trade and distribution information or to add your name to our mailing list and receive a **FREE CATALOG** contact:

Iconografix, Inc.
PO Box 446, Dept BK
Hudson, WI, 54016

Telephone: (715) 381-9755,
(800) 289-3504 (USA),
Fax: (715) 381-9756
info@iconografixinc.com
www.iconografixinc.com

More great books from Iconografix

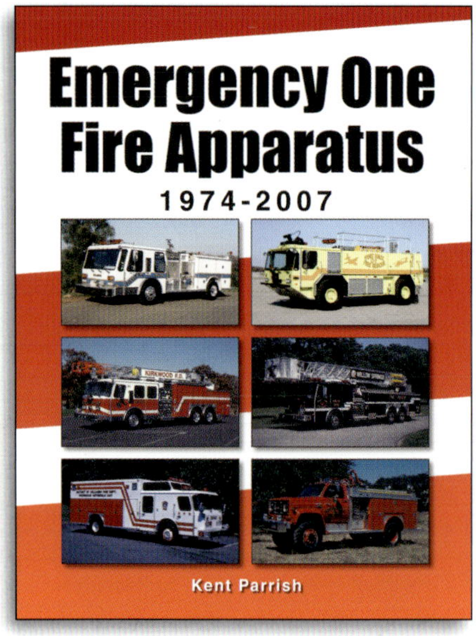

ISBN 978-1-58388-225-2

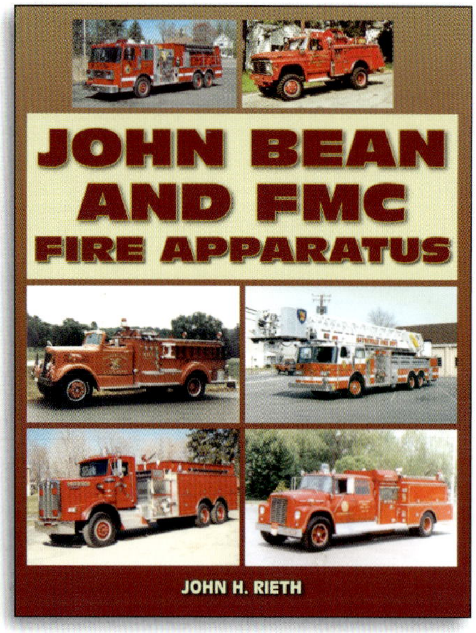

ISBN 978-1-58388-237-5

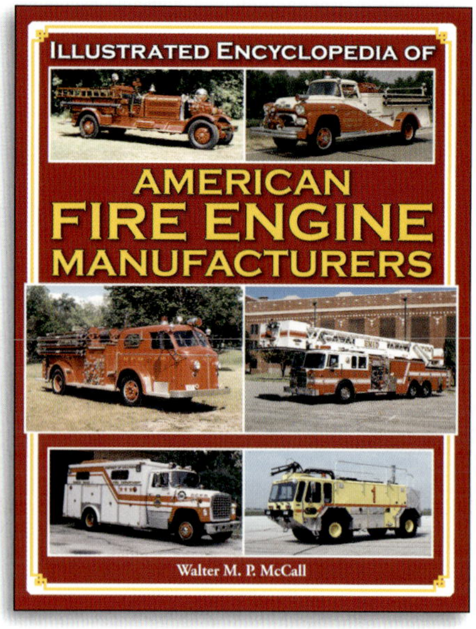

ISBN 978-1-58388-203-0

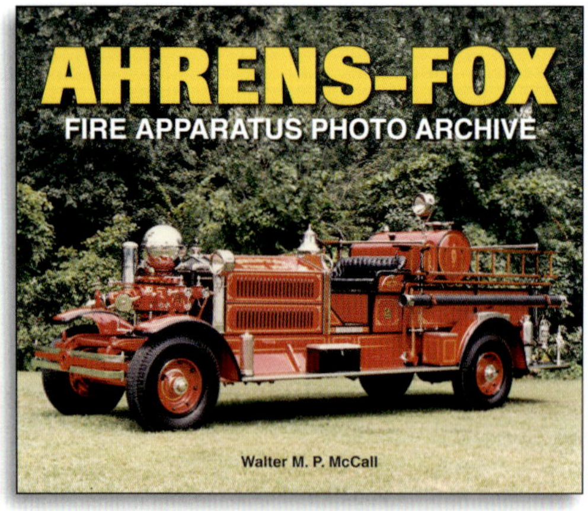

ISBN 978-1-58388-217-7

Iconografix, Inc.
P.O. Box 446, Dept BK,
Hudson, WI 54016
For a free catalog call:
1-800-289-3504
info@iconografixinc.com
www.iconografixinc.com

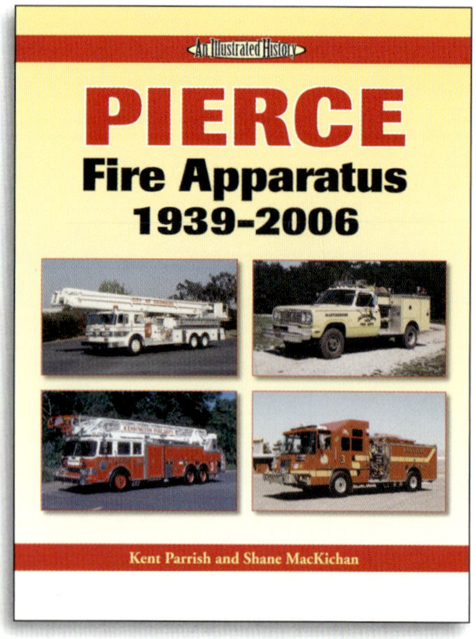

ISBN 978-1-58388-189-7

ISBN 978-1-58388-210-8

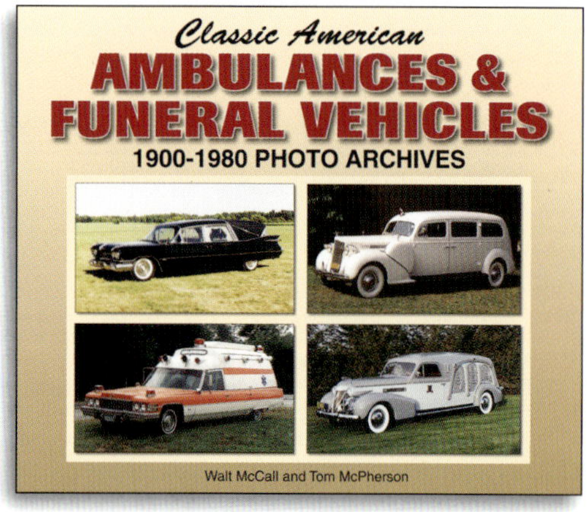

ISBN 978-1-58388-206-1